# HYBRID WARFARE
*The Changing Character of Conflict*

# HYBRID WARFARE
*The Changing Character of Conflict*

*Editor*

Vikrant Deshpande

INSTITUTE FOR DEFENCE STUDIES & ANALYSES
NEW DELHI

**Hybrid Warfare: The Changing Character of Conflict**
*Editor: Vikrant Deshpande*

First Published in 2018

Copyright © Institute for Defence Studies and Analyses, New Delhi

ISBN 978-93-86618-35-1

All rights reserved. No part of this publication may be reproduced, stored in a retrieval system, or transmitted, in any form or by any means, electronic, mechanical, photocopying, recording, or otherwise, without first obtaining written permission of the copyright owner.

**Disclaimer:** The views expressed in this book are those of the authors and do not necessarily reflect those of the Institute for Defence Studies and Analyses, or the Government of India.

*Published by*
PENTAGON PRESS
206, Peacock Lane, Shahpur Jat
New Delhi-110049
Phones: 011-64706243, 26491568
Telefax: 011-26490600
email: rajan@pentagonpress.in
website: www.pentagonpress.in

*In association with*
Institute for Defence Studies and Analyses
No. 1, Development Enclave,
New Delhi-110010
Phone: +91-11-26717983
Website: www.idsa.in

Printed at Avantika Printers Private Limited.

# Contents

|  |  |
|---|---|
| Preface | vii |
| Contributors | xi |
| List of Abbreviations | xiii |
| List of Tables and Figures | xvii |

| | | |
|---|---|---|
| 1. | The Changing Character and the Taxonomy of Conflict<br>*Gurmeet Kanwal* | 1 |
| 2. | Contextualising Hybrid Warfare<br>*Vikrant Deshpande and Shibani Mehta* | 25 |
| 3. | Pakistan's Hybrid War in South Asia: Case Study of India and Afghanistan<br>*Vivek Chadha* | 37 |
| 4. | Russia and Hybrid Warfare: Achieving Strategic Goals without Outright Military Force<br>*Aman Saberwal* | 62 |
| 5. | The Mutating Wars: 'The Hybrid Threat' in Iraq and Syria<br>*Shruti Pandalai* | 74 |
| 6. | Lebanon–Yemen Marathon: Hezbollah Head and Houthi Legs<br>*Kishore Kumar Khera* | 99 |
| 7. | Israel and the Challenges of Hybrid Warfare<br>*S. Samuel C. Rajiv* | 122 |
| 8. | Expanding the Turbulent Maritime Periphery: Gray Zone Conflicts with Chinese Characteristics<br>*Abhay K. Singh* | 141 |

| 9. | India and Hybrid Warfare<br>*Alok Deb* | 173 |
| 10. | Conclusion<br>*Neha Kohli* | 185 |
|  | *Index* | 192 |

# Preface

Are conventional wars or 'war as we knew it' over? A scan of recent conflicts, which are characterised by blurring lines between war and peace, state and non-state, regular and irregular, conventional and unconventional, seems to suggest so. An attempt to answer this question in binaries is fraught with complications. A positive proclamation would render a nation vulnerable to conventional attacks while a negative assertion would create the dilemma of resource allocation. How does one develop additional capabilities required for modern conflicts without reducing conventional capabilities with limited resources? The answers, therefore, are not binary but in the gray zone. This book is an exploratory work in that gray zone.

The armed forces the world over, and in India, are faced with a quintessential dilemma in planning forces for the future. Should they be designing their forces for classic conventional inter-state conflict with the flexibility of fighting non-state actors and hybrid wars, or should they primarily prepare for such hybrid conflicts while retaining some core capabilities for conventional wars? The strategic narrative amongst the armed forces has not progressed from the industrial age metrics of warfare which resort to quantifying casualties both in men or material or capturing valuable real estate. This is evident from the way the armed forces have been structured; however, they seem to be preparing for a war they have not been asked to fight while constantly adapting to a conflict which was not mandated as their primary task.

These small wars, or niggling wars as some have called it, have also been termed as hybrid, non-linear, gray zone, unrestricted and a plethora of such names. The ontological and epistemological enquiry of these terms is essential to understand if they allude to the same phenomenon through different frames. Are they the convention or an aberration? This book tries to fill this crucial research gap in the strategic discourse in India related to the changing character of conflicts. It is inspired by the moot question raised by Brigadier Gurmeet Kanwal during the deliberations over this project about the kind of wars/conflicts the nation will have to fight in the future.

The Military Affairs Centre at the Institute for Defence Studies and Analyses (IDSA) has the right blend of serving/retired armed forces officers and scholars, necessary to deal with the complexities of such an issue. The book covers a wide array of subjects related to international relations and theory of wars and conflicts as well as their contemporary prosecution. The Centre, while conceptualising this project was confronted with two approaches to hybrid warfare. The first was to study each component of hybrid conflicts independently and analyse them for their relevance in the Indian context. The second was to use the case studies model in order to deconstruct modern conflicts. The latter was chosen as it was felt that hybridity cannot be studied as isolated components. Each component gets calibrated as per its own and other's successes and hence, a comprehensive approach to identify and contextualise various components used throughout the world would be apt.

The authors have made an attempt to identify various components of hybrid warfare at play in contemporary conflicts. The work does not intend to be judgmental of the nation states involved in these conflicts or their context and purpose. Instead, the contributors have viewed these conflicts through the prism of the changing character of war, as students of defence and strategic studies to draw relevant lessons.

This volume is divided into three parts. The first part dwells on conceptual issues and contains two chapters. The first chapter by Gurmeet Kanwal deals with the changing character of conflict, while the second chapter by Vikrant Deshpande and Shibani Mehta explores hybrid warfare and similar constructs to arrive at a common understanding.

The second part of the book has six case studies. The first case not only resonates with the Indian reader but also with the scholars themselves. Pakistan has been accused of waging a proxy war against India and Afghanistan. In this study, Vivek Chadha closely examines the various components at play in the South Asian context with Pakistan as a perpetrator of hybrid warfare. Russia appears to have understood, conceptualised and applied hybrid warfare before the rest of the world and the second study by Aman Saberwal deals with Russia and its application of non-linear methodologies in Crimea and Ukraine. West Asia has been riddled with conflicts in the post-colonial era. It has also been a testing ground for power play by other nations post the Cold War. Hybridity and complexity define conflicts in this region, with a host of state and non-state actors acting in concert with external players with *casus belli* ranging from ideological or religious differences to colonial legacies. The majority of the case studies in this book try to deconstruct the complexities of these conflicts in West Asia. Shruti Pandalai covers Syria and Iraq, Kishore Kumar Khera draws a parallel between Yemen and Lebanon from a hybrid perspective and Samuel Rajiv looks at Israel and its use of components of hybrid warfare. The last case study by Abhay Singh is a detailed analysis of China and its use of hybrid devices.

In the last part, the book concludes with two chapters. A chapter by Alok Deb about the lessons learnt from hybrid conflicts in the Indian context makes some recommendations on the way ahead both in terms of policy/strategy as well as capability development and a concluding chapter by Neha Kohli summarizes all issues discussed in the book.

A cursory literature review indicates that while a lot has been written on contemporary conflicts and there are collected essays on hybrid warfare, such an approach of scanning the geopolitical space to deconstruct modern conflicts and draw out lessons learnt is unique. The authors hope to initiate a discourse and a debate on the way India will have to fight its conflicts in the future in order to shape its armed forces and be better prepared for the next Trojan horse.

16 February 2018 **Vikrant Deshpande**

# Contributors

**Brigadier Gurmeet Kanwal (Retd)** is a Distinguished Fellow at the IDSA and an Adjunct Fellow, CSIS Washington DC.

**Vikrant Deshpande** is a serving Army officer and is currently a Research Fellow at IDSA.

**Shibani Mehta** is a Junior Programme Associate at the Takshashila Institution, Bangalore and was an intern at IDSA.

**Colonel Vivek Chadha (Retd)** is a Research Fellow at the IDSA and his research focus is on defence and counter terrorism.

**Aman Saberwal** is a serving officer of the Indian Navy and is currently a Research Fellow at IDSA.

**Shruti Pandalai** is an Associate Fellow at the Military Centre at IDSA, primarily working on issues related to India's national security and foreign policy.

**Kishore Kumar Khera** is a serving fighter pilot of the Indian Air Force and is currently a Research Fellow at IDSA.

**S. Samuel C. Rajiv** is an Associate Fellow at IDSA.

**Abhay K Singh** is a former surface warfare officer of Indian Navy and is a Research Fellow at IDSA.

**Alok Deb** is a retired Army officer and currently Deputy Director General of the IDSA.

**Neha Kohli** is the Associate Editor of the *Journal of Defence Studies*, published by the IDSA, New Delhi.

# List of Abbreviations

| | |
|---|---|
| AAA | Anti-Aircraft Artillery |
| ADIZ | Air Defence Identification Zone |
| AEW&C | Airborne Early Warning and Control |
| AGPL | Actual Ground Position Line |
| AIFV | Armoured Infantry Fighting Vehicle |
| AMD | Anti-Missile Defence |
| APCs | Armoured Personnel Carriers |
| APEC | Asia-Pacific Economic Cooperation |
| AQAP | Al Qaeda in the Arabian Peninsula |
| ASEAN | Association of Southeast Asian Nations |
| ASMs | Anti-Ship Missiles |
| ATGM | Anti-Tank Guided Missile |
| ATMs | Automated Teller Machines |
| AVC | Bureau of Arms Control, Verification and Compliance |
| BATs | Border Action Teams |
| CAPF | Central Armed Police Forces |
| CAR | Conflict Armament Research |
| CERT | Computer Emergency Response Team |
| CIA | Central Intelligence Agency |
| CMC | Central Military Commission |
| COMINT | Communications Intelligence |
| COTS | Commercial Off-the-Shelf |
| CPC | Communist Party of China |
| CPEC | China–Pakistan Economic Corridor |
| CRBN | Chemical, Biological, Radiological and Nuclear |
| DDOS | Distributed Denial-of-Service |

| | |
|---|---|
| EEZ | Exclusive Economic Zone |
| ELINT | Electronic Intelligence |
| EU | European Union |
| 4GW | Fourth Generation Warfare |
| FICN | Fake Indian Currency Notes |
| FIF | Falah-e-Insaniat Foundation |
| FSA | Free Syrian Army |
| GCC | Gulf Cooperation Council |
| GDP | Gross Domestic Product |
| GoM | Group of Ministers |
| HIC | High Intensity Conflict |
| HM | Hizbul Mujahideen |
| IAEA | International Atomic Energy Agency |
| ICVs | Infantry Combat Vehicles |
| IDF | Israel Defense Forces (IDF) |
| IDSA | Institute for Defence Studies and Analysis |
| IEDs | Improvised Explosive Devices |
| IISS | International Institute for Strategic Studies |
| ISAB | International Security Advisory Board |
| ISI | Inter-Services Intelligence |
| ISIS | Islamic State of Iraq and Syria |
| J&K | Jammu and Kashmir |
| JCG | Japanese Coast Guard |
| JCOs | Junior Commissioned Officers |
| JCPOA | Joint Comprehensive Plan of Action |
| JKLF | Jammu and Kashmir Liberation Front |
| JuD | Jamaat-ud-Dawa |
| kg | kilogram |
| km | kilometre |
| LeT | Lashkar-e-Taiba |
| LIC | Low Intensity Conflict |
| LoC | Line of Control |
| Lt Gen | Lieutenant General |
| MAC | Multi Agency Centre |
| MANPADS | Man-Portable Air Defence Systems |
| MBTs | Main Battle Tanks |
| MFA | Ministry of Foreign Affairs |

| | |
|---|---|
| MIC | Medium Intensity Conflict |
| NATGRID | National Intelligence Grid |
| NATO | North Atlantic Treaty Organization |
| NCTC | National Counter Terrorism Centre |
| NGOs | Non-Governmental Organisations |
| NIA | National Investigation Agency |
| NM | Nautical Miles |
| NSA | National Security Agency |
| NSAG | Non-State Armed Group |
| NVGs | Night Vision Goggles |
| NWNP | No War–No Peace |
| ICAO | International Civil Aviation Organization |
| OIC | Organisation of Islamic Cooperation |
| OOTW | Operations Other than War |
| PA | Palestinian Authority |
| PLA | People's Liberation Army |
| PLAN | PLA Navy |
| PLFP | Popular Front for the Liberation of Palestine |
| PLO | Palestine Liberation Organization |
| PoK | Pakistan occupied Kashmir |
| PoWs | Prisoners of War |
| PPP | Purchasing Power Parity |
| PRC | Popular Resistance Committees |
| SAARC | South Asian Association for Regional Cooperation |
| SAGWs | Surface-to-Air Guided Weapons |
| SALW | Small Arms and Light Weapons |
| SCO | Shanghai Cooperation Organisation |
| SDF | Syrian Democratic Forces |
| SIGINT | Signals Intelligence |
| SNC | Syrian National Council |
| SOF | Special Operations Forces |
| SSMs | Surface-to-Surface Missiles |
| TPP | Trans-Pacific Partnership |
| TV | Television |
| UAE | United Arab Emirates |
| UAVs | Unmanned Aerial Vehicles |
| UN | United Nations |

| | |
|---|---|
| UNCLOS | United Nations Convention on the Law of the Sea |
| UNIFIL | United Nations Interim Force in Lebanon |
| UNSC | United Nations Security Council |
| US | United States |
| USSR | Union of Soviet Socialist Republics |
| VBIEDs | Vehicle-Borne IEDs |
| VPNs | Virtual Private Networks |
| WHO | World Health Organization |
| WMD | Weapons of Mass Destruction |

# List of Tables and Figures

## Tables

| | | |
|---|---|---|
| Table 1.1: | Number of Wars, 1946–20027 | 6 |
| Table 2.1: | Components of Hybrid Warfare | 34 |
| Table 3.1: | Components of Hybrid Warfare Employed by Pakistan | 57 |
| Table 5.1: | Hybrid Wars: Variables, Impact, Response | 79 |
| Table 5.2: | Snapshot of Actors and Conflicting Interests in Syria's Hybrid Wars | 86 |
| Table 8A.1: | Gray Zone Incidents in East China Sea and South China Sea | 168 |

## Figures

| | | |
|---|---|---|
| Figure 6.1: | Pictorial Depiction of Warring Factions and External Support Elements in Yemen | 106 |
| Figure 8.1: | Number of Intrusions of Chinese Government and Other Vessels into Japan's Territorial Sea around Senkaku | 152 |
| Figure 8.2: | Number of Scrambles by Japan against Airspace Intrusion by China | 153 |

# 1

# The Changing Character and the Taxonomy of Conflict

*Gurmeet Kanwal*

---

*The categories of warfare are blurring and no longer fit into neat, tidy boxes. One can expect to see more tools and tactics of destruction—from the sophisticated to the simple—being employed simultaneously in hybrid and more complex forms of warfare.*
—Robert M. Gates, Former US Secretary of Defence[1]

## Era of Strategic Uncertainty

The twentieth century was mired in conflict and was arguably the bloodiest in history. The period of the Cold War due to the North Atlantic Treaty Organization (NATO) and the Warsaw Pact, was one marked particularly by strife and chaos. Numerous wars and small conflicts the world over resulted in loss of life and property on a very large scale. One or the other great power was invariably behind these conflicts, directly or indirectly. The peace dividend that was expected to accrue at the end of the Cold War failed to materialise. The unexpected and sudden break-up of the Soviet Union created a power vacuum and further exacerbated the prevailing uncertainties. Long-suppressed ethno-nationalist aspirations for autonomy and self-governance came to the fore the world over. Fissiparous tendencies surfaced where the existence of fissures in society could never have even been imagined—for example, in former Yugoslavia. Movements for democracy came to the fore in many countries governed by dictatorial and authoritarian regimes.

The break-up of the Soviet Union resulted in the United States (US) becoming the sole superpower. However, it could exult in what Charles Krauthammer called the 'unipolar moment'[2] for just about a decade. The 11 September 2001 terrorist attacks on the World Trade Center twin towers and the Pentagon brought about a paradigm shift in the emerging world order. For the first time in history, nation-states became vulnerable to unpredictable threats and non-state actors suddenly became a force to reckon with. In those stunning acts of international terrorism, the new century, and indeed the new millennium, witnessed the dawn of an era of strategic uncertainty. Samuel Huntington's hypothesis about the 'clash of civilisations',[3] scoffed at by analysts as hyperbole only a decade earlier, suddenly seemed ominously realistic.

Today, the global system is caught up in revolutionary upheaval. The concept of the nation-state, the most basic building block of the global system, is itself changing. Approximately one-third of all the present members of the United Nations (UN) are threatened by ethnic disharmony, rebel movements and insurgencies. National borders are becoming increasingly porous; currency rates keep going out of control of the central banks; imports and immigrants are moving freely across the world; and terrorists, guns and drugs are threatening the sovereignty of nations. Out of this chaos, a new kind of political entity, being described as the 'post-national' state,[4] is emerging. It is imperative that the intricate nuances of the various aspects of the changes taking place in the international order, and their repercussions on national security as well as on the political, socio-economic, cultural and ideological components of society, are understood in the correct perspective, so as to formulate meaningful strategies for the future progress, development, well-being and survival of mankind.

## Changing Character of Conflict

The nature of conflict does not change; it endures over a long period of time. However, the character of conflict evolves with the passage of time. Particularly since the end of the Cold War, there has been immense change in the character of conflict. A balance of power system, tentative and skewed as it was, has ensured that the world has so far been spared the spectre of the Third World War, which will inevitably see the use of nuclear weapons. Perhaps history is now working in reverse: the possession of nuclear weapons has ensured that the days of large-scale inter-state wars are almost over and, in their place, the world is witnessing the rise of 'crummy little wars'[5]—fought by insurgents, terrorists, guerrillas, bandits, drug cartels and criminal networks.

The prevailing security environment is radically different from what it was even a decade ago. The probability of conventional conflict between states or groups of states has been steadily declining while, at the same time, sub-conventional conflict is gaining prominence. 'Wars of interest' were supplemented by 'wars of conscience' as the international community, newly awakened to the horrors of the violation of human rights, moved to relieve the suffering of those who had been long oppressed and those who were being victimised for sectarian and ethnic differences. To these two categories of 'wars of interest' and 'wars of conscience', a new category, 'wars of intervention', has been added and military intervention is being justified on many grounds. Non-state actors with transnational presence are emerging as important entities and are gaining prominence that is (almost always) disproportionate to their size and status.

## Dilution in the Concept of the Nation-State

The changing character of conflict is indirectly influencing the conceptualisation of national security in the twenty-first century. The concept of the Westphalian nation-state[6] has begun to gradually fray at the edges. In the post-Cold War world order, power blocs have been slowly giving way to cooperative regional groupings like the European Union (EU)—though Brexit has dented EU cohesion—and trade blocks like Asia-Pacific Economic Cooperation (APEC) and the Trans-Pacific Partnership (TPP). While regional groupings such as the South Asian Association for Regional Cooperation (SAARC), the Association of Southeast Asian Nations (ASEAN) and the Shanghai Cooperation Organisation (SCO) enjoy the advantage of proximity and cultural understanding, these do not possess suitable operational capability in the domain of stability and peace support operations. Regional groupings have been better at issuing declarations and identifying principles than at formulating concrete operational policies for reasons of conflicting national priorities and domestic political considerations. This is not unusual, given that most of them were founded on the mandate of trade and development and have begun exploring security cooperation only recently.

Along with West Asia, Southern Asia has gradually emerged as one of the key epicentres of conflict and instability in the world. Being host to a mix of indigenous peoples and migrants, Southern Asia has witnessed the movement of people for several centuries and many Southern Asian states have rarely seen true political unity. Territorial disputes, religious fundamentalism, radical extremism, ethnic tensions and socio-economic disparities are the hallmarks of Southern Asia. The shadow of nuclear weapons has also contributed to instability in Southern Asia, though no political advantage has been gained by any of the countries possessing these weapons in this conflict-ridden region. At present, it appears unlikely that a genuinely cooperative security framework will eventually emerge in the Indo-Pacific from the ashes of the ongoing conflicts.

## Emerging Contours of Conflict

In the increasingly globalised world, the emerging security challenges are no longer products merely of conventional inter-state rivalries but of economic, demographic and societal tensions that are transnational in nature. The incidence of conflict is on the rise due to multiple factors ranging from weak and illegitimate state institutions, marginalisation of people in border areas (generating sanctuaries for various insurgent groups), large-scale population displacements to in effective regional security arrangements. Modern conflict is more likely to be a consequence of regional struggles involving a range of actors rather than inter-state tensions. Instability is likely to arise as a consequence of the rise of autonomous armed groups and non-state entities and the weakening of governments and state institutions, coupled with population displacement, trafficking—both human and material—and ethno-religious tensions. In some cases, non-state actors act as proxies for inimical nation-states.

Given the rising importance of cities as political, economic and cultural centres of gravity, the battlefields of armed conflict are increasingly shifting towards urban settings. An emerging phenomenon that is gradually gaining momentum is the use of the techniques of information warfare, organised crime and acts of terrorism, fostered by cross-border linkages between disparate terrorist organisations, involving military training, funding and transfer of technology. Cybersecurity is posing new challenges and nation-states are finding it difficult to cope with the increasingly sophisticated hacking techniques being employed by non-state actors and rogue individuals. Non-contact warfare, like economic measures designed to harm a country's economic stability—for example, through the circulation of fake currency—will add to the challenges to be overcome by security planners.

The rising competition over limited energy resources is generating new tensions in geopolitical relations. Its adverse impact is being felt increasingly in the Southern Asian region as well. Future water wars are already being spoken of in hushed tones as a distinct possibility. Though trade wars are in the realm of speculation at present, with increasing economic competition in future, these may not be far off. However, in the foreseeable future, asymmetric, amorphous, cross-cultural conflict will continue to dominate the strategic landscape. It is the rise of these and other non-traditional security threats that will influence both domestic and international policy in the years ahead. While these concerns have been part and parcel of human existence for many years, never before have they had such a serious impact on individual states or the international community as a whole.

These changes in the character of conflict are leading to the gradual transformation of military forces. A nation's armed forces were formerly designed primarily for conventional state versus state conflict. These are now being reoriented to be able to fight a conventional war as well to act decisively against non-state adversaries. As future threats and challenges are becoming increasingly more difficult to predict due to strategic uncertainty, in areas that are devoid of territorial disputes, the force transformation trendline will be to move from threat-based to capability-based forces. Similarly, training regimes will need to be configured to train for certainty and educate for uncertainty.

Conflict continues to be commonplace with no clear distinction between war and peace. While the number of civil wars has been increasing since the end of World War 2, there has been a marked decrease in the number of inter-state wars (See Table 1.1). Even relatively minor conflicts that are localised sometimes have major implications for world peace and stability and often hamper trade and commerce. This makes it important for many nations that are not directly engaged in a conflict to intercede so as to monitor, manage and resolve actual or potential conflicts. Modern conflicts are often marked by increasingly stark asymmetries between the contenders. On one side is usually the state with well-equipped, modern forces but limited public support and severe political and moral constraints. On the other side are irregular combatants organised into small groups of lightly armed forces, with utter disdain for international law, total commitment to the cause, scant regard for life and property and often with overwhelming public support.

**Table 1.1: Number of Wars, 1946–2002[7]**

*Source*: Department of Peace and Conflict Research at Uppsala University; and International Peace Research Institute, Oslo.

## The Taxonomy of Modern Conflict

Modern conflict goes by many names; the different categories are separated by various shades of gray. Possibly the only element common to all types of conflict is violence. 'Violence' has been defined by the World Health Organization (WHO) as 'the intentional use of physical force or power, threatened or actual, against oneself, another person, or against a group or community, which either results in or has a high likelihood of resulting in injury, death, psychological harm, maldevelopment, or deprivation.'[8] The term violence also connotes an aggressive tendency to act out destructive behaviour. Violence falls into essentially two forms: random violence, which includes unpremeditated or small-scale violence; and coordinated violence, which includes actions carried out by sanctioned or unsanctioned violent groups as in war (that is, inter-societal violence) and terrorism.

The *British Defence Doctrine* defines 'civil war' as one that is 'conducted largely within the boundaries of a state in which a significant part of the population is associated with opposing sides. The contest is for government of the state or regional autonomy or secession. One or both sides may have external help.'[9] According to Colonel Gabriel Bonnet, 'revolutionary warfare' 'consists in the application of irregular warfare methods to the propagation of an ideology or political system.'[10] 'Irregular warfare' denotes a form of conflict where one or more protagonists adopt irregular methods.

Irregular troops are combatants who are not formally enlisted in the armed forces of a nation-state or other legally constituted entity.

The two terms revolutionary and irregular warfare, often used interchangeably, owe much to the theories of Mao Tse-tung, Vo Nguyen Giap, Che Guevara and Carlos Marighela. Revolutionary or irregular warfare usually relies on guerrilla tactics that are best summarised in Mao Tse-tung's celebrated remarks:

> Divide our forces to arouse the masses, concentrate our forces to deal with the enemy. The enemy advances, we retreat; the enemy camps, we harass; the enemy tires, we attack; the enemy retreats, we pursue...These tactics are just like casting a net; at any moment we should be able to cast it or draw it in. We cast it wide to win over the masses and draw it in to deal with the enemy...[11]

The word 'militant' has come to refer to any individual or party engaged in aggressive physical or verbal combat, normally for a cause.[12] Militant is an often-used neutral term for soldiers who do not belong to an established military. Typically, a militant engages in violence as part of a claimed struggle for achievement of a political goal. Popular usage sometimes sees 'militants' as synonymous with terrorists. The term 'militant state' colloquially refers to a state that holds an aggressive posture in support of an ideology or cause. The term militant also describes those who aggressively and violently promote a political philosophy in the name of a movement (and sometimes have an extreme solution for their goal). The present phase of militancy in Jammu and Kashmir (J&K), which originated in 1988–1989, had a clearly stated initial goal to gain *azadi* or independence from India.

Frank Kitson has defined 'insurgency' as the 'use of armed force by a section of the people against the government to overthrow those governing the country at the time or to force them to do things which they do not want to do.'[13] The British Army defines insurgency as an:

> organised movement aimed at the overthrow of a constituted government through the use of subversion and armed conflict...Some insurgencies aim to seize power through revolution. Others attempt to break away from state control and establish an autonomous state within ethnic or religious boundaries...Generally an insurgent group attempts to force political change by a mix of subversion, propaganda, political and military pressure.[14]

Only well-organised insurgencies with a strong leadership and widespread popular support are capable of posing a viable long-term threat to the state.

Successful insurgencies tend to have external political, diplomatic and military support, including for training and logistics, sanctuary in the supporting nation, the ability to control some territory and, at the culminating stage, the ability to raise well-trained and motivated battalions that are almost at par with the opposing army.

'Counter-insurgency' is defined by the British Army as 'Military, paramilitary, political, economic, psychological and civic actions taken by a government to defeat insurgency.'[15] Counter-insurgency is characterised by relatively infrequent combat at section, platoon and company levels rather than formation level, with consequently a lower rate of logistics consumption than in major combat. However, counter-insurgency campaigns typically continue for several years.

Insurgencies often have transnational linkages and hence, a successful counter-insurgency campaign must fight and break the links between insurgencies across a wide region to prevent recurrence. The British Army doctrine lists the following principles for fighting such a campaign:[16]

- Ensure political primacy and political aim.
- Build coordinated government machinery.
- Develop intelligence and information.
- Separate the insurgent from his support.
- Neutralise the insurgent.
- Plan for the long term.

The term 'fourth generation'[17] warfare (4GW) has come to characterise post-modern conflict in which operations are conducted in a decentralised manner and where the distinction between war and politics and that between combatant and civilian is blurred. Also, the nation-state is losing its so-far unchallenged monopoly on the use of military force. The term 4GW represents any war in which one of the opponents is a non-state actor and not a duly constituted nation-state. Present-day conflict is thought to be more akin to that than conflict in pre-modern times.

## Gray Zone Conflict

*This is another type of war, new in its intensity, ancient in its origin— war by guerrillas, subversives, insurgents, assassins, war by ambush instead of by combat; by infiltration, instead of aggression, seeking victory by eroding and exhausting the enemy instead of engaging him.*

—John F. Kennedy[18]

Another term that is gaining in currency is gray zone conflict. It refers to a 'gray zone' between conventional and sub-conventional conflict.

According to the US Special Forces Command white paper, gray zone challenges are:

> defined as competitive interactions among and within state and non-state actors that fall between the traditional war and peace duality. They are characterized by ambiguity about the nature of the conflict, opacity of the parties involved, or uncertainty about the relevant policy and legal frameworks.[19]

A US International Security Advisory Board (ISAB) report, commissioned by the Bureau of Arms Control, Verification and Compliance (AVC) at the Department of State, states, 'What makes this concept particularly relevant today is both the greater extent to which these tactics are being utilised by our adversaries, and also the expanded technological tool kit that can be brought to bear in these efforts.'[20] The report makes six major recommendations:[21]

- Taking a whole-of-government approach to countering gray zone efforts, including engaging agencies whose primary focus is not international security.
- Setting up an organisational structure for gray zone operations that will facilitate coordination and management across the full range of US government agencies engaged.
- A renewed focus on planning to face these challenges, including a sober assessment of the US goals, objectives and interests in different countries and regions around the world.
- Continuing the US efforts to address the fundamental underlying sources of violence and the conditions that make gray zone tactics potentially effective—by promoting economic opportunity, justice, human rights, good governance, public health and the rule of law.
- Developing, both at the Department of State and across the US government, a better 'after action/lessons learned' system, as well as an additional focus on training and 'wargaming' gray zone scenarios with stakeholders across government.

The Special Operations Command white paper concludes:

> We should seek to identify, understand, and highlight activities running counter to US interests. This awareness can help attribute nefarious activity, potentially increasing costs for that activity even if the US does directly intervene...The US already has most of the tools required to secure and advance its national security interests in the gray zone. However, it must evolve its organizational, intellectual

and institutional models to flourish in the middle ground between war and peace and avoid the predictability and rigidity characterizing its actions since the end of the Cold War.[22]

## Age of 'New Terrorism'

Terrorism, the latest scourge that has infested geopolitics with violence organised by both state and non-state actors, has defied definition because of its complexity. The UN Secretary-General's High-level Panel on Threats, Challenges and Change has described terrorism as:

> any action, in addition to actions already specified by the existing conventions on aspects of terrorism, the Geneva Conventions and Security Council resolution 1566 (2004), that is intended to cause death or serious bodily harm to civilians or non-combatants, when the purpose of such an act, by its nature or context, is to intimidate a population, or to compel a Government or an international organisation to do or to abstain from doing any act.[23]

The High-level Panel's definition is wide-ranging and short on specifics. According to the *British Defence Doctrine*:

> Terrorism may be defined as premeditated, politically motivated violence perpetrated by groups or individuals and usually intended to influence an audience wider than that of its immediate victims. In one form, terrorism may be an element of insurgency. In another, it may be employed for objectives short of the overthrow of the state. It may also be used by one state against another.[24]

The advent of the Islamic State of Iraq and Syria (ISIS) and the brutal brand of fundamentalist terrorism practised by the ISIS militia mark the coming of age of the era of 'new terrorism'. It hit India with the Mumbai serial bomb attacks of March 1993. In the same year, a group of Islamist extremists led by Ramzi Yousef launched the first attack on the World Trade Center in New York. In 1995, Aum Shinrikyo attacked the Tokyo underground with Sarin gas. Soon after that, a large truck bomb killed 168 people in Oklahoma City and visions of apocalypse through terrorism began to haunt the world. The London and Madrid train bombings further heightened the pervasive fear psychosis. Peter R. Neumann, a journalist, academic and commentator on terrorism and political violence, quotes Walter Laqueur, the well-known terrorism historian, as having

> noted that a 'revolution' in the character of terrorism was taking place. Rather than the vicious yet calculated application of violence that everyone had become familiar with, the world was now confronted

with terrorists whose aim was 'to liquidate all satanic forces [and destroy] all life on earth'.[25]

The 11 September 2001 attacks were a catastrophic confirmation of a major shift in the trend lines of transnational terrorism and the multiple terror strikes in Mumbai in November 2008 provided further proof of a new form of terrorism. There is now ready agreement that the age of 'new terrorism' is well and truly upon us. However, in many ways, 'new terrorism' is still a catch phrase that heralds change as no clear understanding of its characteristics is as yet forthcoming. Even as the world attempts to enhance its understanding of what exactly has changed, four key patterns can be clearly discerned.

First, modern terrorist organisations are both diffuse and opaque in nature. They have cellular structures that resemble networks, rather than a clearly demarcated chain of command. Second, they are increasingly more transnational in their geographical spread, with shifting centres of gravity and constantly changing recruitment bases. Third, their ideological motivations are driven by religious fundamentalism and they seek to achieve their political objectives through radical extremism even though no religion justifies violent means. Fourth, modern terrorism is far more violent than 'old' terrorism. In the mid-to-late twentieth century, it used to be said that terrorist organisations wanted 'a lot of people watching, not a lot of people dead'. In the last two decades this has changed and they now wish to inflict horrendous casualties so that they can impose their will on governments and societies.

Neumann has written:

> Regardless of whether governments are dealing with 'old' or 'new', the aim must be to prevent terrorist attacks whilst maintaining legitimacy in the eyes of the population. In doing so, governments need to 'harden' potential targets; develop good intelligence in order to disrupt terrorist structures; bring to bear the full force of the law whilst acting within the law; address legitimate grievances where they can be addressed; and, not least, convey a sense of calm and determination when communicating with the public.[26]

This prescription cannot be faulted and policy planners across the world would do well to draw up a counter-terrorism policy on these lines as part of a comprehensive national security strategy.

## Small Wars

The term 'small war' is a literal translation of the Spanish word *guerrilla*. This term was popular around 1900 to refer to encounters between Western troops and irregular or guerrilla forces in the Third World. The US has a long but largely uncelebrated history of fighting 'small wars', and 'if the past is a prologue of what is to come, small wars will be the main occupation of the American military for the foreseeable future', says Max Boot.[27] According to the US Marine Corps manual on small wars:

> As applied to the United States, small wars are operations undertaken under executive authority, wherein military force is combined with diplomatic pressure in the internal or external affairs of another state whose government is unstable, inadequate, or unsatisfactory for the preservation of life and of such interests as are determined by the foreign policy of our Nation.

Small wars were primarily inter-state conflicts fought to achieve foreign policy objectives even though these were often skewed. The *Small Wars Journal* states on its homepage:[28]

> We believe that Small Wars are an enduring feature of modern politics...The characteristics of Small Wars have evolved since the Banana Wars and Gunboat Diplomacy. War is never purely military, but today's Small Wars are even less pure with the greater interconnectedness of the 21st century. Their conduct typically involves the projection and employment of the full spectrum of national and coalition power by a broad community of practitioners.

> 'Small Wars' is an imperfect term used to describe a broad spectrum of spirited continuation of politics by other means, falling somewhere in the middle bit of the continuum between feisty diplomacy and global thermonuclear war...

> The term 'Small War' either encompasses or overlaps with a number of familiar terms such as counterinsurgency, foreign internal defense, support and stability operations, peacemaking, peacekeeping, and many flavors of intervention. Operations such as noncombatant evacuation, disaster relief, and humanitarian assistance will often either be a part of a Small War, or have a Small Wars feel to them. Small Wars involve a wide spectrum of specialized tactical, technical, social, and cultural skills and expertise, requiring great ingenuity from their practitioners.

Small wars are likely to be small for four reasons.[29] First, the political objectives of military intervention are likely to be specific. Second, finite political objectives will tend to limit the military objectives. Third, limited military objectives and the political necessity to keep the scope of the conflict as non-threatening to other states as possible restrains nations from bringing to bear all the force they have available. Last, they are small because the likely enemies may be unable to engage in anything larger than a small war unless other countries sustain them. If other countries do sustain them, thus compelling an increase in the forces to secure original objectives or new and larger ones, warfare may escalate from the small category into something else. Nonetheless, the size or site of the conflict may not always be a good pre-conflict indicator of its intensity.

Intensity is the product of many interactive variables, including the value placed on objectives, the strength of the opposed wills and the armaments and training of the forces engaged. In his philosophy on warfighting, codified in *Fleet Marine Forces Manual 1*, the Commandant of the Marine Corps, General Alfred M. Gray, asserts that intensity is determined by the "density of fighting forces or combat power on the battlefield..." However, not all small wars can be so classified if the number of casualties is taken as the yardstick of measurement, as the following figures show:[30]

- Taiping Rebellion, 1851–1854: 2 million dead.
- US Civil War, 1861–1865: 800,000 dead.
- Great War in La Plata, 1865–1870: a million dead.
- Sequel to the Bolshevik Revolution, 1918–1920: 600,000 dead.
- First Chinese–Communist War, 1927–1936: a million dead.
- Spanish Civil War, 1936–1939: 2 million dead.
- Communal riots in the Indian Peninsula, 1946–1948: 800,000 dead.

## Low Intensity Conflict (LIC)

One method of classifying conflict is based on its intensity. In this method, conflict is classified as low intensity conflict (LIC), medium intensity conflict (MIC) and high intensity conflict (HIC). An LIC may be 'non-violent' (subversion, show of force, peacekeeping under Chapter VI of the UN Charter) or 'violent' (revolutionary or guerrilla war, counter-insurgency, terrorism, prolonged confrontation along a line of control [LoC], peacekeeping under Chapter VII of the UN Charter). The LIC is a generic term that is prevalent in intra-state wars and is rarely used in the context of inter-state wars. Inter-state wars, such as the Arab–Israeli wars of 1967 and 1973 and the Indo-Pak wars of 1947–48, 1965 and 1971, with higher levels of violence and consequently a larger number of casualties than LIC, are usually classified as MIC as these are generally short of full-scale all-out conventional conflict. A full-fledged conventional war like the First

World War and the Second World War would normally be referred to as HIC. If there were ever a nuclear war in future, it would naturally fall in the HIC category.

The US Department of Defense defines LIC as:

> Limited politico-military struggle to achieve political, social, economic or psychological objectives. It is often protracted and ranges from diplomatic, economic and psychological pressure through terrorism and insurgency. Low intensity conflict is generally confined to a geographical area and is often characterised by constraints on weaponry, tactics and the level of violence.[31]

The Indian Army's definition of LIC is more wide-ranging:

> LIC is a generic term encompassing all kinds of armed conflicts that are above the level of peaceful co-existence among states and below the threshold of war. These include proxy war, terrorism and insurgencies; border skirmishes also fall within this category. It involves protracted struggle of competing principles and ideologies.[32]

An LIC is characterised by one or all of the following conditions:

- Asymmetry of force levels between the regular forces and the irregular opposition force.
- The force applied and the violence generated depends on the code of conduct and the capabilities of the weaker side.
- Laws of the land impose restrictions on the actions of the security forces.[33]

Some thinkers aver:

> while LIC is theoretically possible in a modern industrial nation, it is a form of conflict most appropriate to the Third World. Furthermore, it can be stated that this concept can be applied only in cases where there is no direct confrontation between the superpowers since such a confrontation, should armed conflict actually commence, could scarcely be stabilized at a LIC level. Although allied, friendly or client regimes of either side or one of the superpowers themselves may be involved, LIC theory does not allow for direct conflict between the United States and the Soviet Union in the Third World.[34]

Not all analysts agree with the all-encompassing definition of LIC. There are many detractors of the term LIC who are of the view that such categorisation will hamper understanding of an emerging war form rather than enhance it (in fact, many in the US are more comfortable with the term Operations Other than War [OOTW]). Some opposing views are as follows:

- Lieutenant Colonel John Fulton of the School of Advanced Military Studies argues that in creating LIC, 'the doctrine community may be creating a doctrinal foster home for orphaned warfare concepts...LIC's definition is too broad, and the category is too large.'[35]
- Colonel Dennis Drew of Air University found it to be a 'dismally poor title for a type of warfare in which thousands die, countless more are physically or psychologically maimed and, in the process, the fate of nations hangs in the balance.'[36]
- General John R. Galvin has stated, 'The simple classification into high and low intensity conflict can be dangerous if it inhibits our understanding of what the fighting is all about.'[37]

William Olsen of the US Army War College states:

In actual fact, the definition of LIC should not concentrate on the military level of conflict, but on its political character...The aim is not military conquest, but social control, for whose attainment military means can be employed as an element of the struggle...The use of military force must be measured by its social and political utility. Military means are a tactical element of a strategic program that emphasises goals and means. Though important, the use of military might is limited, while the use of diplomatic and political means may be unlimited.[38]

It emerges that LIC is a concept that is not of a purely military nature, even though it has been developed and propounded chiefly by various militaries the world over. An LIC requires an integrated politico-economic–military approach, supplemented by psychological, social and diplomatic support. It can be stated without exaggeration that, conceptually, a successful counter-LIC campaign primarily requires a politically oriented integrated policy approach containing essential military elements—it is not first and foremost a military matter. Also, it is often forgotten that for a soldier at the receiving end of an insurgent's machine gun fire, the intensity is anything but low.

## Hybrid Conflict

As the evolving threats are asymmetric in nature, the concept of 'hybrid warfare' is becoming increasingly more mainstream. The terms 'hybrid warfare', 'hybrid conflict', 'hybrid war' or 'hybrid threat' have become common in international armed forces and academic literature. While there is no universally accepted definition of hybrid warfare, the following definition is quite comprehensive:

> Hybrid conflicts...are full spectrum wars with both physical and conceptual dimensions: the former, a struggle against an armed enemy and the latter, a wider struggle for, control and support of the combat zone's indigenous population, the support of the home fronts of the intervening nations, and the support of the international community...To secure and stabilise the indigenous population, the intervening forces must immediately rebuild or restore security, essential services, local government, self-defence forces and essential elements of the economy.[39]

A 'hybrid threat' has been described as: 'Any adversary that simultaneously and adaptively employs a tailored mix of conventional, irregular, terrorism and criminal means or activities in the operational battlespace. Rather than a single entity, a hybrid threat or challenger may be comprised of a combination of state and non-state actors.'[40]

Clearly, in hybrid conflict, the challenge faced by a nation's armed forces goes beyond simply a military challenge. The Hezbollah's tactics and practices during the 2006 Lebanon conflict and the ISIS militia's early campaigns against the governing regimes in Iraq and Syria exemplify the intricacies of hybrid conflict. The political, social, diplomatic and informational components that act as force multipliers make the Hezbollah and the ISIS militia far more potent than the power of their military capability. An effective response to hybrid warfare requires a comprehensive or what is being increasingly called 'whole-of-government' approach.

Some analysts are of the view that 'melding the capabilities of irregulars, political indoctrination, assassination, regular military forces, and diplomacy' is better described as 'compound warfare' rather than hybrid conflict. The examples given are those of the British campaigns in Malaya (Malaysia) and Northern Ireland, as also the resistance offered by the North Vietnamese to the US forces. Frank Hoffman describes compound warfare as:

> ...those major wars that had significant regular and irregular components fighting simultaneously under unified direction... Compound wars offered synergy and combinations at the strategic level, but not the complexity, fusion, and simultaneity we anticipate at the operational and even tactical levels in wars where one or both

sides is blending and fusing the full range of methods and modes of conflict into the battlespace. Irregular forces in cases of compound wars operated largely as a distraction or economy of force measure in a separate theatre or adjacent operating area including the rear echelon. Because it is based on operationally separate forces, the compound concept did not capture the merger or blurring modes of war identified in past case studies such as Hizbollah in the second Lebanon war of 2006...[41]

Commenting on Hoffman's description of compound warfare, Glenn has written:

With his mention of 'full range of methods and modes of conflict,' Hoffman lends further support to the argument that the broader, more-than-military challenge posed by Hizbollah and similar threats is worthy of further intellectual pursuit...The issue is not one of whether the comprehensive approach and whole of government constructs also apply at this (tactical) level—they undoubtedly do— but rather whether the nature of operations at the tactical level such as those approaches employed by Hizbollah constitute a form of warfare unique from conventional and irregular operations. It is certainly possible that while a hybrid concept may prove un-unique at the operational and strategic levels; its tactics constitute a different kind of fighting.[42]

The Russians prefer to use the term 'non-linear' warfare, exemplified by their intervention in Crimea and eastern Ukraine. Confronted with what they perceive as an unfriendly if not hostile world order, the Russians are increasingly relying on new tactics that focus on the weaknesses of the adversary rather than engaging him in direct combat. Conventional as well as irregular military forces are employed in conjunction with diplomatic, economic, cyberwarfare and psychological attacks, which heighten feelings of insecurity in the target population. It is also being referred to as the Gerasimov Doctrine. Dichotomies in the existing international law are being particularly exploited: for example, whether or not cyberwarfare, electronic warfare and information warfare are acts of war. Recent international experience calls for a re-examination of the legal framework defining the acts that constitute aggression. Hybrid conflict is analysed in greater detail in the next chapter.

Two Chinese Colonels focused the world's attention on yet another variation of hybrid conflict, which they called 'unrestricted warfare', when they wrote a book by this name.[43] Literally, the term means 'warfare beyond bounds'. Finding China in a David versus Goliath type of asymmetric military situation, Qiao Liang and Wang Xiangsui came to the conclusion that irregular warfare would be more effective than direct combat. They argue that there are now several forms of warfare that can be practised: 'diplomatic, financial, network, trade, bio-chemical, intelligence, resources, ecological, psychological, economic aid, space, tactical, regulatory, electronic, smuggling, sanction, guerrilla, drug, news media, terrorist, virtual, ideological warfare, and many more.' The term signifies strategies that militarily (and politically) disadvantaged nations might adopt in order to counter more powerful opponents. Though it is not revolutionary, the term has caught the imagination of analysts the world over as a modern-day version of the use of all the elements of national power to achieve the desired objectives, as had been advocated by Clausewitz in his model of 'total war'. The authors do not explain how a legitimately installed government can implement indiscriminate, perhaps illegal, means to achieve its goals.

## Sub-conventional Conflict in the Indian Context

Independent India has been embroiled in conflicts of one variety or the other. The long-drawn, so-called 'eyeball-to-eyeball' confrontation along the LoC with Pakistan since the first war over J&K in 1947–48 and along the Actual Ground Position Line (AGPL) at Siachen Glacier in northern Kashmir since April 1984 can only be described as a 'low intensity limited war'. The intervening periods of relative peace, such as the present period of the informal ceasefire that has been in force since 25 November 2003, are referred to as period of no war–no peace (NWNP) in the Indian Army circles.

In the early 1980s, Pakistan supported Sikh militancy in Punjab with a view to encourage some disaffected Sikhs led by Jarnail Singh Bhindranwale to establish an independent state that was to be called Khalistan. However, the movement did not have a mass base and eventually, the Indian Army, the Punjab Police and the people of Punjab got together in the mid-1990s to defeat Pakistan's diabolical machinations. Since 1988–1989, Pakistan has been waging what has been called a 'proxy war' against India in J&K and elsewhere through its mercenary marauders, the so-called 'jihadis', who are armed, equipped, trained and financed by the Pakistan Army and the Inter-Services Intelligence (ISI) Directorate. According to the *Indian Army Doctrine*:

> Proxy war is a war conducted between nations utilising non-state players to fight on their behalf. At least one of them employs a third

party to fight on its behalf. The extent and type of support provided by the states involved in proxy war will vary but financial and logistics support are always provided.[44]

Throughout the Cold War, the two superpowers fought proxy wars all over the world. This was done by one superpower providing military, diplomatic and financial aid to its surrogates to enable them to fight adversaries supported by the other superpower. At any one time, 10–20 such conflicts were being fought across the world. Since Pakistan's proxy war against India does not materially concern the major powers, it has not got the attention it deserves.

Another modern-day scourge afflicting India and much of the rest of the world is terrorism. According to the *Indian Army Doctrine*:

> Terrorism is the unlawful use of force or violence against people or property to terrorise, coerce or intimidate governments or societies. This is most often resorted to with the aim of achieving political, religious or ideological objectives. Terrorism thrives on a fear psychosis and could be achieved by arson, sabotage, hijacking, hoaxes, maiming, bombing, seizure, kidnapping, assassination, taking hostages, raids, ambushes and the use or threat of use of WMD.[45]

Since the 11 September 2001 terrorist strikes against targets in New York and Washington, DC, terrorist attacks in India have been receiving some attention. Though it has taken long, the West has finally realised that the source of terrorism against India and the West is the same. Kashmiri terrorists are no longer being called freedom fighters. Another inappropriate distinction that was being made—and still is in many cases—by the media is that those who launch an attack on civilians in Western cities are called terrorists and those who do so in the Third World are called gunmen.

## Concluding Observations

No matter which term it goes by—civil disobedience, counter-insurgency, guerrilla warfare, insurgency, insurrection, internal security, revolutionary warfare, small wars, subversion, terrorism, 4GW, gray zone or hybrid—conflict in the first quarter of the twenty-first century is predominantly sub-conventional conflict that is more often intra-state than inter-state. The root causes of modern conflict are primarily socio-economic and ethno-religious tensions that transcend state boundaries, rather than territorial and boundary disputes. It is mostly a contest between state and non-state actors, and often a triangular one between disparate groups of non-state actors and the state. There are seldom any victors but the vanquished abound in the shape of displaced and homeless persons, those who are severely wounded or maimed for life, those who cannot find productive employment and those who spend prolonged periods under custody, sometimes without even a charge sheet being filed.

Most people's image of what a war is like is shaped by what they see on their television screens. When people think of war, they conjure up images of blood-and-guts wars like the two World Wars. They think of mass mobilisation, of conscription, of major disruptions in civilian life and of body bags being brought home. When civilians in positions of authority think of war, they think of conventional conflicts. Their preferred style of war is usually the Second World War or the Gulf War. They do not like small wars and LICs, which do not have clear-cut outcomes, drag on endlessly, do not have exit strategies and force troops to act as social workers. Commanders in army headquarters the world over dislike such conflicts even more as these demand unconventional responses that dilute finely honed command and control systems and result in handing over the charge of conflicts to company commanders, subalterns and sergeant majors or, on the Indian sub-continent, to Junior Commissioned Officers (JCOs). The Powell Doctrine, which held that America should only fight if it is going to use overwhelming force, win a massive victory and then leave immediately, soon fell by the wayside as the US was deeply, almost inextricably, involved in long-drawn sub-conventional conflict in both Afghanistan and Iraq.

It emerges clearly that while the probability of state-on-state conflict has definitely declined, its possibility cannot be ruled out. Conventional conflict in future will be marked by greater violence and devastation, with a high probability that nuclear weapons may be employed on the battlefield. It will have a deep impact on national economies and international trade, and major international effort will be necessary for damage clearance and reconstruction.

Not all analysts agree that modern conflicts can be explained or understood in new terms. A contrary view merits mention:

> The argument advanced here seeks to demonstrate that terms like 'guerrilla warfare' and 'low intensity conflict' are fundamentally flawed. They do not exist as proper categories of war. Often they constitute inappropriate distinctions that impede intellectual understanding of internal war phenomena, which has in the past had a negative impact upon policymaking. The usage of these terms

in strategic studies literature does not facilitate understanding but rather undermines the attempt to comprehend the complexity of warfare as a whole. What we call low intensity conflict can be fully understood—can only be understood—within Clausewitzian parameters, which embrace the entire spectrum of war.[46]

Martin van Creveld has written:

> The roughly three-hundred-year period in which war was associated primarily with the type of political organisation known as the state—first in Europe, and then, with its expansion, in other parts of the globe as well—seems to be coming to an end. If the last fifty years or so provide any guide, future wars will be overwhelmingly of the type known, however inaccurately, as 'low intensity'. Both organisationally and in terms of the equipment at their disposal, the armed forces of the world will have to adjust themselves to this situation by changing their doctrine, doing away with much of their heavy equipment and becoming more like the police. In many places that process is already well under way.[47]

With regard to 'the lessons of history', Boot offers this advice:

> In deploying American power, decision makers should be less apologetic, less hesitant, less humble. Yes, there is a danger of imperial overstretch and hubris—but there is an equal, if not greater, danger of under commitment and lack of confidence. America should not be afraid to fight 'the savage wars of peace' if necessary to enlarge the empire of liberty. It has done it before.[48]

Perhaps Indian decision makers too ought to heed this advice when it comes to considering hard policy options for intervention in India's neighbourhood.

Perceptive observers of world politics disagree about the approaching outlook for war.[49] Is the world in the midst of an era of peace with a declining prospect of war, or is it facing a future characterised by increasing small wars and LIC driven by long-suppressed ethnic tensions, religious fundamentalism, socio-economic inequities and a revolution of rising expectations? This puzzle will continue to drive social scientists to strive for a more comprehensive examination of the phenomenon of conflict. The changes that have come about in the character of conflict require nation-states to revisit their understanding of the strategic, legal and psychological facets of warfare and modify their policies and plans in accordance with the changes.

## NOTES

1. Robert M. Gates, 'A Balanced Strategy: Reprogramming the Pentagon for a New Age', *Foreign Affairs*, January–February 2009, available at https://www.foreignaffairs.com/articles/united-states/2009-01-01/balanced-strategy, accessed on 06 September 2017.
2. Charles Krauthammer, 'The Unipolar Moment', *Foreign Affairs*, Winter 1990–1991 available at https://www.foreignaffairs.com/articles/1991-02-01/unipolar-moment, accessed on 05 February 2018.
3. Samuel P. Huntington, 'The Clash of Civilizations?', *Foreign Affairs*, Summer 1993, available at https://www.foreignaffairs.com/articles/united-states/1993-06-01/clash-civilizations, accessed on 05 February 2018.
4. A post-national state may be described as one that has no core identity, no mainstream. Justin Trudeau, the Canadian Prime Minister, has described his country as a post-national state. Charles Foran, 'The Canada Experiment: Is this the World's First "Postnational" Country?', *The Guardian*, 4 January 2017, available at https://www.theguardian.com/world/2017/jan/04/the-canada-experiment-is-this-the-worlds-first-postnational-country, accessed on 06 September 2017.
5. Martin van Creveld, *The Transformation of War*, New York: The Free Press, 1991.
6. The treaties of Westphalia, signed in 1648, laid the foundation of the modern nation-state system and of the concept of territorial sovereignty.
7. UN, *A More Secure World: Our Shared Responsibility*, Report of the UN Secretary-General's High-level Panel on Threats, Challenges and Change, 2004, p. 11, available at file:///C:/Users/Admin/Documents/UN-AMoreSecureWld-SharedRespo-UNSG-Report-2004.pdf, accessed on 05 February 2018.
8. Etienne G. Krug, Linda L. Dahlberg, James A. Mercy, Anthony B. Zwi and Rafael Lozano (eds), *World Report on Violence and Health*, Geneva: WHO, 2002.
9. 'War and Armed Conflict: The Dimensions of Conflict', in *British Defence Doctrine*, Joint Warfare Publication (JWP) 0-01, London: Secretary of State for Defence, 1996, p. 2.15.
10. Vivek Chadha, *Company Commander in Low Intensity Conflict*, New Delhi: Lancer Publishers and Distributors, 2002, p. 12.
11. 'A Single Spark can Start a Prairie Fire', in *Selected Writings of Mao Tse-tung*, Peking, 1968.
12. 'Difference between Militant and Terrorist', A Reportersite, 18 June 2016, available at https://areportersite.wordpress.com/2016/06/18/difference-between-militant-and-terrorist/, accessed on 06 September 2017.
13. Frank Kitson, *Low Intensity Operations*, cited in Chadha, *Company Commander in Low Intensity Conflict*, n. 10, p. 10.
14. British Army, *Army Doctrine Publication: Land Operations*, Directorate General Development and Doctrine, London, May 2005, p. 17.
15. Ibid.
16. Ibid., p. 18.
17. The concept of 'generations' of warfare is usually attributed to William S. Lind and his associates. William S. Lind, Keith Nightengale, John F. Schmitt, Joseph W. Sutton and Gary I. Wilson, 'The Changing Face of War: Into the Fourth Generation', *Marine Corps Gazette*, October 1989, pp. 22–26, available at http://

18. John F. Kennedy, 'Remarks at West Point to the Graduating Class of the U.S. Military Academy, 06 June 1962', in Gerhard Peters and John T. Woolley (compilers), *The American Presidency Project*, available at http://www.presidency.ucsb.edu/ws/?pld=8695.
19. The US Special Operations Command, 'The Gray Zone', White Paper, 09 September 2015, p. 1, available at https://info.publicintelligence.net/USSOCOM-GrayZones.pdf accessed on 06 September 2017.
20. ISAB, *Report on Gray Zone Conflict*, Washington, DC: AVC, US Department of State, 03 January 2017, available at https://www.state.gov/t/avc/isab/266650.htm, accessed on 06 September 2017.
21. Ibid.
22. The US Special Operations Command, 'The Gray Zone', White Paper, 09 September 2015, n. 19.
23. UN, *A More Secure World*, n. 7, p. 49.
24. British Army, *Army Doctrine Publication: Land Operations*, n. 14.
25. Peter R. Neumann, 'Old and New Terrorism', Social Europe, 03 August 2009, available at https://www.socialeurope.eu/old-and-new-terrorism, accessed on 05 February 2018.
26. Ibid.
27. Max Boot, *The Savage Wars of Peace: Small Wars and the Rise of American Power*, New York: Basic Books, 2002.
28. *Small Wars Journal*, available at http://smallwarsjournal.com/content/about, accessed on 05 February 2018.
29. Richard Szafranski, 'Thinking about Small Wars', *Parameters,* September 1990, pp. 39–49, available at http://carlisle-www.army.mil/usawc/parameters/1990/szafran.htm, accessed on September 6, 2017.
30. Lewis F. Richardson, 'The Statistics of Deadly Quarrels', available at http://world.std.com/~jlr/comment/statistics.htm, accessed on September 6, 2017.
31. James J. Gallagher, *Low Intensity Operations,* p. 3, cited in Chadha, *Company Commander in Low Intensity Conflict*, n. 10, p. 11.
32. 'Types of Wars', in *Indian Army Doctrine*, Shimla: HQ Army Training Command, October 2004, p. 16, Chapter 2, Section 5, available at www.ids.nic.in/Indian%Army%Doctrine/indianarmydoctrine_1.doc, accessed on 05 February 2018.
33. Ibid.
34. Jochen Hippler, 'Low Intensity Warfare and its Implications for NATO', December 1988, available at the author's website: http://jochenhippler.de/html/low_intensity_warfare_and_its_implications_for_nato_-_1988.html, accessed on 05 February 2018.
35. John S. Fulton, 'The Debate about Low-Intensity Conflict', *Military Review*, Vol. 66, February 1986, p. 61.
36. Dennis M. Drew, 'Insurgency and Counterinsurgency: American Dilemmas and Doctrinal Proposals', Center for Aerospace Doctrine, Research and Education, USAF Air University, September 1987, p. 5.
37. John R. Galvin, ''Uncomfortable Wars: Toward a New Paradigm', *Parameters*, Winter 1986, Vol. 16, No 4, pp 2-8.
38. William J. Olson, 'Airpower in Low-intensity Conflict in the Middle East', Ninth

Air University Airpower Symposium on the Role of Airpower in Low-intensity Conflict, Maxwell AFB, Alabama, 11–13 March 1985, pp. 218–221.
39. John J. McCuen, 'Hybrid Wars', *Military Review*, March–April 2008, pp. 107–113, cited in Dr. Russell W. Glenn, "Thoughts on Hybrid Conflict", *Small Wars Journal*, available at http://smallwarsjournal.com/blog/journal/docs-temp/188-glenn.pdf, accessed on 05 February 2018.
40. Definition adopted at the US Joint Forces Command Hybrid War Conference held in Washington, DC, 24 February 2009, cited in Russell W. Glenn, 'Thoughts on "Hybrid" Conflict', *Small Wars Journal*, 4 March 2009, available at http://smallwarsjournal.com/blog/journal/docs-temp/188-glenn.pdf, accessed on 06 September 2017.
41. Frank G. Hoffman, 'Hybrid Warfare and Challenges', *Joint Forces Quarterly*, Issue 52, 1st Quarter, 2009, pp. 34–38. The article provides a useful articulation of hybrid warfare and related concepts.
42. Glenn, n. 39.
43. Qiao Liang and Wang Xiangsui, *Unrestricted Warfare*, Beijing: PLA Literature and Arts Publishing House, 1999.
44. *Indian Army Doctrine*, n. 32.
45. Ibid.
46. M.L.R. Smith, 'Guerrillas in the Mist: Re-assessing Strategy and Low Intensity Warfare', *Cambridge Journals Online*, Vol. 29, No. 1, 2003, available at http://www.journals.cambridge.org/action/displayAbstract?fromPage=online&aid=135275, accessed on 06 September 2017.
47. Martin van Creveld, *Through a Glass, Darkly: Some Reflections on the Future of War*, 2000, available at http://www.d-n-i.net/creveld/through_%20a_glass_darkly.htm.
48. Boot, *The Savage Wars of Peace*, n. 27, accessed on September 6, 2017.
49. M.R. Sarkees, F.W. Wayman and J.D. Singer, 'Inter-State, Intra-State, and Extra-State Wars: A Comprehensive Look at their Distribution over Time, 1816–1997', *International Studies Quarterly*, Vol. 47, No. 1, March 2003, pp. 49–70.

# 2

# Contextualising Hybrid Warfare

*Vikrant Deshpande and Shibani Mehta*

War (or conflict) is primarily a clash of interests. A violent struggle between 'two hostile, independent and irreconcilable wills, each trying to impose itself on the other'.[1] The argument about the evolving character of conflict made in the previous chapter is best exemplified by the ends, ways and means construct which defines strategy.[2] The end has always been to influence the adversary's will in varying degrees, from achieving political advantage to decisive victory. The means (resources available) have been used in innovative ways (courses of action) according to the prevailing political and social landscape and advances in science and technology. Swords, annexation through matrimonial alliances and establishing control over the indigenous people for commercial purposes have made way for airstrikes, information campaigns and energy diplomacy, but the struggle for power has remained consistent through the ages. Modern form of rivalry, statecraft and warcraft fortify classic stratagems with a range of unconventional techniques to achieve politico-strategic goals. The resulting heterogeneity has led to the formulation of a confusing array of names: gray zone, hybrid, gradualist, non-linear, unrestricted, full spectrum and many more.[3] Such unconventional approaches have a renewed relevance in modern conflicts due to new effective tools that are being extensively used by the actors involved. This chapter aims to survey these constructs regarding the changing character of war and conflicts and bring analytical coherence to the issue.

The contention is that these approaches, a mix of conventional and unconventional, using all elements of power and political influence, are not entirely new. States have been using these compound or hybrid approaches for centuries. Concepts such as political destabilisation, support for proxies and militias, information campaigns and more have been a staple of state and warcraft as early as the Peloponnesian wars. This blending finds historic examples in the Battle of Troy where a Trojan horse was used as a method of surprise and deception; 'in the Napoleonic Wars where the British regulars challenged French control of major Spanish cities, while Spanish guerrillas attacked their lines of communication; and in the Arab Revolt where the British Army combined conventional operations in Palestine with irregular forces under British operational control.'[4] Closer home, the use of conventional force along with application of unconventional forces is characteristic in the conflict over Kashmir. How are modern conflicts then distinct if the same combination is employed? We intend to present the argument that while previously unconventional methods were used as force multipliers to amplify the application of conventional force, modern conflicts seem to have the case in reverse. An attempt has also been made to explore the causes for such a reversal.

## Characteristics of Conflicts in the Twenty-first Century

Before mapping conflicts of the twenty-first century, it is essential to distinguish between war and conflict. The terms tend to be used interchangeably despite having distinct connotations. War, according to Merriam-Webster, is 'a state of usually open and declared armed hostile conflict between states or nations'. The same dictionary defines conflict as 'competitive or opposing action of incompatibles; antagonistic state or action; a conflict of principles'. These distinctions can be summarised as:

- War is conflict but not all conflicts are wars.
- Conflict is a broad term covering a wide spectrum of competition, ranging from verbal disagreement to the use of force.

In the context of this study, of the compound nature of wars and conflicts across a spectrum of ways and means of coercion, the term conflict is most suitable.

General Rupert Smith opens the introduction to his book, *The Utility of Force*, by addressing the durability of conflict: 'War no longer exists. Confrontation, conflict and combat undoubtedly exist all around the world....'[5] It is this enduring nature of conflicts in the twenty-first century that distinguish them from those of the previous centuries. A Russian General, Valery Gerasimov, provides the most appropriate characterisation of such conflicts in his doctrinal document by stating, 'Wars are no longer declared and, having begun, proceed according to an unfamiliar template.'[6] A survey of modern conflicts reveals blurring lines between not only war and peace but also amongst elements of national power, kinetic and non-kinetic operations, covert and overt acts, policy, strategy, operations and tactics, law and order and public order, diplomacy and warfare, enemy and population, allies and adversaries, virtual and real battlespaces, ethics and morals, state and non-state actors, conventional and unconventional means, and even national boundaries. The age of information has somehow merged or diffused these distinctions. Conventional response mechanisms of the state, hitherto modelled on these distinctions, therefore find it difficult to counter these threats and have to constantly adjust and adapt.

These conflicts seem to fall between war and peace, and therefore have been referred to as gray zone conflicts. The International Security Advisory Board (ISAB) of the Department of State of the United States (US) carried out an exercise to list characteristics of gray zone conflicts and found that it includes the following:

- Cyber, information operations, efforts to undermine public/allied/local/regional resistance, and information/propaganda in support of other hybrid instruments;
- Covert operations under state control, espionage, infiltration and subversion;
- Special Operations Forces (SOF) and other state-controlled armed units, and unacknowledged military personnel;
- Support—logistical, political, and financial—for insurgent and terrorist movements;
- Enlistment of non-governmental actors, including organized criminal groups, terrorists, and extremist political, religious, and ethnic or sectarian organizations;
- Assistance to irregular military and paramilitary forces;
- Economic pressures that go beyond normal economic competition;
- Manipulation and discrediting of democratic institutions, including electoral system and the judiciary;
- Calculated ambiguity, use of covert/unacknowledged operations, and deception and denial; and
- Explicit or implicit threat use, or threats of use of armed force, terrorism and abuse of civilian populations and of escalation.[7]

Both the Russian and American characterisations attempt to deconstruct modern conflicts.

The notable distinction between medieval and industrial age conflicts and modern conflicts is not the combination or hybridity of the various means utilised, but the tools made available by the prevailing social, economic, political and technological environments.

## Effects of Nuclear and Information Age on Modern Conflicts

The technological revolution has resulted in a shift from an economy based on traditional industry to one based on information and automation. This has transformed the social landscape of human life. 'Information is no longer the monopoly of those in power. Ideas can be everywhere at once, in real time and full colour'.[8] The pervasiveness of information technology in human activity has introduced a new relationship between economy, state and society. In an era of round-the-clock news coverage, the population, aided by smartphones, television and social media, can track their government and military with startling frequency. People are now more connected but can also be more mercurial.[9] Powerful imagery like that of body bags coming home, especially in a democracy, can lead to major shifts in war policies. While this holds true across time, the real-time streaming of violence increases the likelihood of this consequence.

It cannot be denied that the information age has provided new ways of wielding coercive influence but attributing the characteristics of modern conflict to advancements made in the information age alone would be a simplistic correlation of little value. The coming of the information age has also altered the determinants of national power, including polity, military, economy, society, infrastructure, culture, ideology and others. Politically, the information age has established a new standard of transparency resulting in a modification in the norms of governance. While *jus ad bellum* has been relevant historically, the information age amplifies it further as states have to justify their actions to domestic and international audiences. This imposes certain restrictions on the use of force, the scale dependent on whether the force is utilised at home or abroad. This allows a weaker adversary to blunt superior manpower and technology by merging with the people. The non-state actor or terrorist organisation has no bounds to its behaviour like the state and utilises the tools of the information age to its advantage.

The merging of cybernetics infrastructure with critical infrastructure has eliminated the need for physical contact and enabled disgruntled citizens to mobilise across geographies and create non-state entities through social and other media. It also provides access to low-cost options for the non-state actor to conduct violent, and sometimes indiscriminate, acts which have disproportionate effects or tactical actions having a direct politico-strategic consequence. It took one train bombing in Spain, three days before a general election, for the withdrawal of Spanish forces from Iraq. This has also resulted in a moral, ethical and legal dilemma regarding acts of terrorism. Conventional responses are inadequate, while unconventional methods such as Guantanamo Bay detention camp rattle the foundations of jurisprudence and society. The non-state actors operate in this gray zone exploiting these dilemmas to the hilt.

The causality of modern conflicts in the context of the information age would require an entire book by itself. It would suffice for this study to conclude that there is a correlation between the two. Some of the causes for mutations of modern conflicts are the effects of the information revolution and others may not be directly connected with the information revolution but are consequences of the changing socio-political dynamics in the modern world. The nuclear dimension, for example, has added an element of unpredictability to the outcomes of conventional conflicts, making them less lucarative in achieving strategic objectives. This, in turn, gives rise to the stability–instability paradox as Lidell Hart states 'to the extent that the H[hydrogen] bomb reduces the likelihood of full-scale war, it increases the possibility of limited war pursued by widespread local aggression.'[10]

Scholars and practitioners have been trying to contextualise these changes in the way modern conflicts are prosecuted since the turn of the century. Competing ideas from different nations emerge as an attempt to encapsulate these conflicts and their complexities. These are: hybrid warfare and gray zone conflicts from the US; unrestricted warfare from China; and non-linear warfare from Russia. It would be prudent to examine these ideas and relate them to conflicts involving India as well as ancient Indian construct on warfare.

## Hybrid Warfare and Gray Zone Conflicts

Frank Hoffman, in his seminal piece on the rise of hybrid wars,[11] was the first of the block in characterising twenty-first century conflicts as hybrid. His study found that 'conventional', 'irregular' and 'terrorist' are not distinct and isolated challenges with fundamentally different approaches. Instead, they are means employed in combination by the adversary and conducted by both state and non-state actors. Hoffman defined hybrid wars as 'blend of the lethality of state conflict with the fanatical and protracted fervor of irregular war.'[12] In modern wars, forces become blurred into the same force, in the same battlespace, with the regular component acting as a decisive force multiplier for strategic effect.

While Hoffman talked of hybrid warfare as a combination of tactics that are regular, irregular and asymmetric, Margaret Bond took this combination beyond the realm of war and defined it as a combination of national power elements.[13] According to her:

> Hybrid war envisions employment of a comprehensive and highly-nuanced variety of military activities, resources, programs, and applications, tailored to maximize a non-violent, persuasive use of economic and political influence to reform hostile governments, movements, or trends in politically, socially, and economically unstable conditions, characteristic of failing and failed states. It also includes a full range of military intelligence capabilities, nonconventional (including nonlethal) weapons, armaments, support units, and combat equipment, available for instant employment if ever opposition elements of regular forces or irregular insurgents, terrorists, or other non-state actors cross the hostility threshold and constitute a direct threat to or threaten these non-hostile activities.[14]

Bond develops a case for a broader, more proactive application of all elements of national power along a continuum of activities from stability to armed combat, including peaceful humanitarian missions and post-conflict reconstruction and stabilisation. Bond's definition, though more comprehensive due to its inclusion of the non-kinetic elements, is predominantly a state perspective.

Another term that has emerged in the US lexicon is gray zone conflicts. The Department of State, in its ISAB report, explains gray zone as: 'denotes the use of techniques to achieve a nation's goals and frustrate those of its rivals by employing instruments of power—often asymmetric and ambiguous in character—that are not direct use of acknowledged regular military forces.' The distinction between hybrid and gray zone seems to be that gray zone excludes from its ambit conventional or regular military forces. This exclusion is contradictory to its own characterisation of gray zone conflicts which, as seen earlier, includes 'explicit or implicit threat use, or threats of use of armed force...'. The term gray zone, therefore, explains only the asymmetric, irregular or unconventional aspects of the changing character of conflict. Excluding conventional or regular warfare from gray zones would be incorrect because the gray zones exist under a looming threat of use of such force and would turn a lighter or darker shade of gray where such forces are excluded or included for use. We can, therefore, safely conclude that gray zone conflicts are a subset of hybrid conflicts; or gray zone describes an environment between war and peace where hybrid tools of coercive influence ranging from conventional forces to criminal activities are utilised.

## Unrestricted Warfare

Two Chinese Colonels, Qiao Liang and Wang Xiangsui, have understood the strains that globalisation places on states and their conflicts. In an analysis titled, *Unrestricted Warfare*, they have said:

> The great fusion of technologies is impelling the domain of politics, economics, the military, culture, diplomacy and religion to overlap each other. The connection points are ready and the trend towards merging of the various domains is very clear. All of these things are rendering more and more obsolete the idea of confining warfare to the military domain and of using the number of casualties as a means of the intensity of war. In warfare and non-military warfare, which is primarily national and supra national, there is no territory which cannot be surpassed, there are no means that cannot be used, and there is no territory and method which cannot be used in combination.[15]

Similar to Bond, this concept exploits the benefits of combining various domains of national power available and uses them as a means to prosecute war. The Chinese Colonels bring an Oriental and relevant flavour by adding culture and religion to elements of national power. According to the editor's note, Qiao has argued in a subsequent interview that 'the first rule of unrestricted warfare is that there are no rules, with nothing forbidden'.[17] This vision clearly transcends any traditional notions of war.

## Non-linear Warfare

In a comparison of Russian non-linear warfare with hybrid warfare, Tad A. Schnaufer II credits General Gerasimov with the conceptual construct of non-linear warfare.[17] The term, however, first appeared in a short story written by one of President Putin's closest political advisors, Vladislav Surkov, under the pseudonym Nathan Dubovitsky, just a few days before the annexation of Crimea in 2014.

According to Schnaufer, the concept of 'non-linear' describes conflict as one that does not have clear front lines or district friendly/enemy areas. Non-linear warfare relies on the subversion and division of the enemy's social and political structure, allowing the aggressor to do their will by any means, not just brute force. The idea of non-linear warfare comes from the approach that it has no bounds and sometimes functions with limited planning, thereby allowing a state to exploit an opportunity. Schnaufer further argues that non-linear warfare is distinct from hybrid warfare as hybrid is a combination of asymmetric, regular, irregular and terrorist tactics, while non-linear warfare transcends into political, diplomatic, economic and informational domains. However, it can be observed that the tenets of non-linear warfare fulfil Margaret Bond's definition of hybrid warfare: projecting all elements of national power as a continuum of activities across the full spectrum. In their analysis, Oscar Jonsson and Robert Seely describe the Russian form of hybrid warfare as Russian full-spectrum conflict, with kinetic violence, information, economic and energy campaigns and political operations spread across the spectrum.[18] This is, again, primarily a variation of Bond's definition.

## Hybrid Warfare in *Arthashastra*

In his treatise, Kautilya refers to four types of wars:

1. *Mantrayuddha* or War by Counsel: This is the exercise of diplomacy to win wars. It is to be utilised when the king is in a weaker position and engaging in battle is not wise or beneficial.
2. *Prakasayuddha* or Open Warfare: This is a form of conventional warfare which follows all laid down rules of fighting a battle. Open warfare, Kautilya declared, is 'most righteous'. He advised the king that 'When he is superior in troops, and when he is on land suitable to himself, he should engage in an open fight...'.
3. *Kutayuddha* or Concealed Warfare: This form of warfare includes irregular methods, ambushes and treachery in the enemy camp.
4. *Guda Yuddha* or *Tusnimyuddha*; Clandestine/Silent War: This type of war is waged by covert means to achieve the objective without actually waging battle, usually by assassinating the enemy. In such a war, the king not only uses his agents and double agents but also allies, vassals, tribal chiefs and the suborned friends and supporters of the enemy.[19]

Despite being an ancient treatise, the *Arthashastra*'s postulates on types of warfare continue to be extremely relevant to modern concepts and components of hybrid warfare: diplomacy, information operations, conventional warfare and unconventional warfare, including subversion, sabotage, covert operations, assassinations, etc. Kautilya, Roger Boesche argues, considered diplomacy as really a subtle act of war, a series of actions taken to weaken an enemy with an eventual aim to conquer.[20] This Kautilyan convergence between diplomacy and warfare and use of all means of political influence (four *upaya*s)—*sama* (conciliation or diplomacy), *dana* (economic gratification), *bheda* (dissension or information operations) and *danda* (use of force)—to achieve the end state resonates with concepts of use of all elements of national power in hybrid, gray zone, unrestricted and non-linear warfare.

Military strategist and philosopher Sun Tzu has presented a similar argument: '...to fight and conquer in all your battles is not supreme excellence; supreme excellence consists in breaking the enemy's resistance without fighting.'[21] The logical conclusion is that conceptual clarity about hybrid warfare existed in the political and military science of the Orient (Kautilya and Sun Tzu) and is being rediscovered by the Occidental theorists as they are confronted by an enemy unwilling to contest them on their strengths in conventional warfare. Therefore, the difference between unrestricted warfare, non-linear warfare, full-spectrum conflict and hybrid warfare is minimal. All the terms allude to a *combined application of more than one form of the elements of power (national is restrictive as non-state actors may also apply these) in a coordinated, coherent and sometimes simultaneous way to achieve a desired political end state*. These elements of power, however, need an explanation as they are not limited to diplomacy, information, military and economy but also include all elements of hard and soft power, including politics, culture, religion and ideology.

## Components of Hybrid Conflicts

If this understanding of hybrid conflicts is to be applied to modern conflicts, then the pertinent question that arises is: when does a conflict become hybrid? The word hybrid implies a combination or compound of more than two elements of power or components of the spectrum of conflict. These components have been typically characterised as conventional and unconventional or regular and irregular. These distinctions, however, are relative. What is unconventional today may become the convention tomorrow. Similarly, irregular may become regular as its usage increases. These subsets are, therefore, imperfect. Kinetic and non-kinetic, however, are distinct and immutable segregations. Kinetic components would include a spectrum ranging from space weapons, nuclear, biological, chemical options, land forces, naval forces, air forces, special forces, insurgents, terrorists to black operatives carrying out illegal activities, including sabotage, assassinations and violent agitations. Non-kinetic would encompass diplomatic and political actions, information operations including cyber and social media operations, network warfare to include disruption of critical network infrastructure, dissension, subversion, criminal activities, economic warfare including fake currencies, currency devaluation and economic coercion, resource warfare, environmental warfare, ideological warfare, non-violent agitations, etc. The list given in Table 2.1 is neither definitive nor exhaustive, for any and every tool available to attain strategic goals, both kinetic and non-kinetic, will be utilised.

Table 2.1: Components of Hybrid Warfare

| *Kinetic* | *Non-kinetic* |
|---|---|
| Space weapons | Diplomatic |
| Land, sea and air power | Information operations |
| Nuclear, chemical and biological weapons | Dissension |
| Insurgents | Subversion |
| Terrorists | Economic warfare |
| Special forces | Resource wars |
| Black operatives—assassinations and sabotage | Weather wars |
| Violent agitations | Non-violent agitations |
| Criminal activities | |

*Any conflict where two or more components are utilised in concert and with convergence of strategic goals may, therefore, be termed as hybrid.* It is well established that the concept of hybrid (though the term is nascent) existed even in Kautilyan times and kinetic and non-kinetic means have been innovatively used to attain strategic goals throughout history. The key metamorphoses in the twenty-first century conflicts are the tools made available by information technology. These have provided new ways as well as affecting and altering the existing patterns of other means and components of conflicts. In other words, not only does the information revolution provide alternative ways and means but it also alters the existing ways and means to achieve the political ends. An alternative nomenclature therefore for these wars and conflicts could be 'conflicts of the information age'. In this book, the terms conflicts of the information age, hybrid warfare, non-linear warfare, unrestricted warfare and full-spectrum conflict will be used interchangeably to describe contemporary conflicts of the twenty-first century as per the country and the context they allude to.

## Conclusion

To summarise, the contextualisation of hybrid warfare yields the following:

- Hybridity in terms of a combination of tools, both kinetic and non-kinetic, has existed throughout the history of conflict.
- The tools available transmute as per existing social–political conditions and the developments in science and technology.
- Hybrid theory, along with gray zone theory, unrestricted warfare and non-linear theory, is an attempt to deconstruct contemporary conflicts.
- Kautilya's *Arthashastra* and Sun Tzu's treatise are proof that such hybridity, gray zones and non-linear approaches existed throughout the history of Oriental political and military science.
- The distinction, if any, between the medieval and industrial age conflicts is not in the hybrid mix or non-linear approaches but in the tools made available in the information age and the socio-political context of this age, which is obviously distinct from the earlier ones.
- The efforts to deconstruct hybrid or gray zone indicate an application of more than one form of the elements of power (national is restrictive as non-state actors may also apply these) in a coordinated, coherent and sometimes simultaneous way to achieve a desired political end state.

Irrespective of what they are termed, hybrid, gray zone, unrestricted or non-linear, are creditable attempts to deconstruct modern conflicts—a process which is carried forward by this chapter and in turn, by the book. An absolute necessity for scholars and practitioners of defence and strategy is to predict the outlines of future conflicts and to structure, train and prepare state institutions (including the armed forces) to tackle and utilise these implements of the information age.

## NOTES

1. United States (US) Marine Corps, *Warfighting*, available at http://www.marines.mil/Portals/59/Publications/MCDP%201%20Warfighting.pdf accessed on 31 July 2017.
2. A.F. Lykke, Jr, 'Defining Military Strategy', *Military Review*, Vol. 77, No. 1, 1997, p. 1, available at https://www.questia.com/library/journal/1P3-13134302/defining-military-strategy accessed on 31 July 2017.
3. M.J. Mazarr, *Mastering the Gray Zone: Understanding a Changing Era of Conflict*, Carlisle, PA: Strategic Studies Institute and US Army War College Press, 2015, pp. 1–8, available at https://ssi.armywarcollege.edu/pdffiles/PUB1303.pdf accessed on 31 July 2017.

4. Alex Deep, 'Hybrid War: Old Concept, New Techniques', *Small Wars Journal*, 2015, available at http://smallwarsjournal.com/jrnl/art/hybrid-war-old-concept-new-techniques, accessed on 31 July 2017.
5. Rupert Smith, *The Utility of Force: The Art of War in the Modern World*, London: Allen Lane, 2005, p. 3.
6. General Valery Gerasimov, Chief of the General Staff of the Russian Federation, 'The Value of Science in Anticipation: New Challenges Require Rethinking the Forms and Methods of Warfare', Rough translation by Robert Coalson [Facebook note], 21 June 2014, available at https://www.facebook.com/notes/robert-coalson/russian-military-doctrine-article-by-general-valery-gerasimov/10152184862563597/accesed on 01 August 2017.
7. ISAB, *Report on Gray Zone Conflict*, Washington, DC: US Department of State, 2017, p. 2, available at https://www.state.gov/documents/organization/266849.pdf accessed on 13 August 2017.
8. Hank Koehn, *Impact of the Information Revolution*, Center for Media Literacy, 2017, available at http://www.medialit.org/reading-room/impact-information-revolution, accessed on 31 July 2017.
9. S. Besson, 'Sovereignty in Conflict', *European Integration Online Papers (EIoP)*, Vol. 8, No. 15, 2004, available at https://papers.ssrn.com/sol3/papers.cfm?abstract_id=594942.
10. B.H. Lidell Hart, *Deterrence and Defense,* London: Steven and Sons, 1960, p. 23 in Michael Krepon, The Stability-Instability Paradox, Misperception and Escalation Control in South Asia available at https://www.stimson.org/sites/default/files/file-attachments/stability-instability-paradox-south-asia.pdf, accessed on 05 February 2017.
11. Frank G. Hoffman, *Conflict in the 21st Century: The Rise of Hybrid Wars*, Arlington, VA: Potomac Institute for Policy Studies, 2007.
12. Ibid., p. 28.
13. Margaret S. Bond, *Hybrid War: A New Paradigm for Stability Operations in Failing States*. Carlisle Barracks, Carlisle, PA: US Army War College, 2007, available at http://www.comw.org/qdr/fulltext/0703bond.pdf accessed on 12 August 2017.
14. Ibid., p. 4.
15. Qiao Liang and Wang Xiangsui, *Unrestricted Warfare*, Beijing: PLA Literature and Arts Publishing House, 1999, p. 162
16. Ibid., p. xviii.
17. Tad A. Schnaufer II, 'Redefining Hybrid Warfare: Russia's Non-linear War against the West', *Journal of Strategic Security*, Vol. 10, No. 1, 2017, pp. 17–31. Retrieved from http://scholarcommons.usf.edu/cgi/viewcontent.cgi?article=1538&context=jss accessed on 18 Jan 2018.
18. Oscar Jonsson and Robert Seely, 'Russian Full-Spectrum Conflict: An Appraisal after Ukraine', *The Journal of Slavic Military Studies*, Vol. 28, No. 1, 2015, pp. 1–22.
19. L.N. Rangarajan, *Kautilya: The Arthashastra*, Noida, Penguin Books, 1992, p. 636.
20. Roger Boesche, 'Kautilya's *Arthasastra* on War and Diplomacy in Ancient India', *The Journal of Military History*, Vol. 67, No. 1, January 2003, pp. 9–37.
21. Sun Tzu, *The Art of War*, translated by S.B. Griffith, Oxford: Clarendon Press, 1964.

# 3

# Pakistan's Hybrid War in South Asia
## Case Study of India and Afghanistan

*Vivek Chadha*

---

Pakistan remains an enigma for most analysts who have kept a consistent and close watch on its approach towards security issues. Amongst these, the unrelenting bid to seek parity with India and control over the government in Afghanistan, possibly remain its most enduring and persistent security objectives. This becomes apparent when one observes Pakistan's obsessive desire in this regard stretching over a number of decades. Its alignments and realignments with global and regional powers in a bid to achieve these strategic aims are amongst the steps undertaken in the past. An assessment of history provides an overview of Pakistan's strategy. This is especially apparent when seen from the military perspective and further, in terms of the variety of elements within the ambit of hybrid war employed against its adversaries.

This chapter argues that Pakistan, irrespective of its comparative military standing with an adversary, has employed hybrid warfare in a bid to achieve its intended strategic objectives. Pakistan had a lower conventional capability in relation to India. Conversely, it had a far superior military strength when compared to Afghanistan. Despite this, the preference for hybrid war reflects its reliance on this form of warfare against both adversaries. The chapter also concludes that the advent of the information age has only reinforced this conviction by increasing the repertoire of tools in the quiver of Pakistan, given its propensity to employ them, especially against India. In India's context, Pakistan's inability to achieve its strategic objectives when employing conventional force as its primary tool, even as the other sub-conventional constituents were subsidiary in nature, emerges on the basis of the wars fought between the two countries in the twentieth century. This suggests that the employment of hybrid warfare is not a guarantee for success. Conversely, failure in assessing the strengths and weaknesses of an adversary can instead lead to failure to achieve designated objectives, as witnessed in 1947–1948, 1965 and 1999. While the results of the ongoing conflict in Kashmir can only be analysed at a future date, Pakistan's reversal of strategy, with sub-conventional endeavours taking centrestage, is noteworthy. In this context, Christine Fair argues that this relates to Pakistan Army's strategic culture, which employs 'various mobilizations of Islam' and 'tools that Pakistan has developed to manage its varied security challenges'.[1] While this is relevant for both Afghanistan and India, the toolset employed by Pakistan varies, as the following assessments will illustrate.

These assessments, despite covering 70 years of conflict, use the term hybrid warfare, not because this term was prevalent during the earlier periods but more because this form of warfare has been undertaken well before the case studies being analysed. And since the constituents employed by Pakistan fall within the conceptual framework of this publication, hybrid warfare will remain the common thread interlinking the course of events.

## Pakistan's Hybrid War against India

Each war or conflict that Pakistan has waged against India has included tenets of hybrid war. The effectiveness of the same is debatable, as is the success in achieving intended objectives. However, with the advent of the information age, the enhanced toolset available to Pakistan has transformed the means of waging hybrid war. Unlike the previous wars of the industrial age, it has also provided a varying notion of victory, which could relate the continuing failure of India to limit Pakistan's interference in Kashmir as its virtual success, at least in the short term. However, the future course of history could well judge this apparent success differently, if the blowback action of nurturing radicalisation eventually does more damage to Pakistan itself than to India.

A brief assessment of Pakistan's military interventions inside India will illustrate the changing dynamics employed by Pakistan as part of its hybrid strategy and its resultant impact.

## 1947–1948 Indo-Pak War

The first instance of Pakistan's hybrid strategy was witnessed during the 1947–1948 war between the two countries. It is an acknowledged fact that Pakistan's regular forces formally became a part of the war in May 1948.[2] However, what indeed makes the Pakistani effort hybrid was its decision to employ tribesmen, led and supported by retired and serving officers of the army. This was an attempt to characterise the war as an insurrection by the people against their ruler. It was simultaneously aimed to keep the conflict below a threshold, which would avoid an open war between the two countries, a situation that was considered disadvantageous by Pakistan. And this was a framework that they were able to successfully function within, since India did not enlarge the theatre of conflict beyond Jammu and Kashmir (J&K).

The insurrection itself not only involved the well-documented act of plundering and killing of Kashmiris but also subversion of tribals and members of the armed as well as state forces, which led a number of them to revolt against the Maharaja of J&K.[3] This was guided by Sardar Ibrahim Khan, an influential politician who subsequently became the first President of Pakistan occupied Kashmir (PoK). The highest political leadership in the newly created Pakistan was also involved from the very inception of the hybrid conflict. This included Prime Minister Liaquat Ali Khan; the Minister for Refugees, Mian Iftikhar-ud-Din; Brigadier (later Major General) Akbar Khan; and the Finance Minister, Ghulam Mohammad. They were part of a conference, especially convened in September 1947 at Lahore, to coordinate the overall effort in Kashmir.[4]

In line with the appreciation prepared by Akbar Khan, two forces were earmarked for the armed insurrection. One of them was tasked to operate along the Muzaffarabad–Srinagar axis under Major Khurshid Anwar (Retd) and another under Major Zaman Kiani along the southern axis towards Kathua.[5] Akbar Khan was given the responsibility for provision of logistics, by virtue of his appointment as head of the Weapons and Equipment Directorate. However, he was required to execute this plan surreptitiously, without the knowledge of the British military hierarchy, which at that time still led the Pakistan Army.[6]

This strategy did not work as envisaged due to a number of factors. Beyond the initial plan, its follow-up lacked the intimate involvement of Pakistan's military and political hierarchy, especially during the initial phase of operations. It further witnessed poor senior leadership and an over reliance on the ability of tribals to achieve military success against a professional Indian Army. According to Major Amin, a military analyst, the centre of gravity of the entire operation remained on the Jhelum Valley, with Srinagar as its objective. However, the inability to pursue the same, both as a result of constrained vision, weak command and control structures within the Pakistani leadership and resolute defensive actions by Indian military and political leaders, led to the plan's failure.

The Pakistani hierarchy was hesitant to get overtly involved in the conflict since it feared an all-out war between the two countries. This would have been disastrous for Pakistan, given the conventional superiority of Indian forces. The inherent constraint raises the question of viability of pursuing an objective which is unlikely to be achieved, except under the best-case scenario. Unrealistic optimism became evident as a trait in 1947–1948, and this was repeated time and again thereafter as well, as an assessment of wars fought between India and Pakistan indicates. Pakistan's hope of achieving success through a sub-conventional or a limited conflict in J&K in the hope that India will not enlarge the conflict was as flawed then as it was later during subsequent wars. Having said that, the ability of Pakistani strategists to achieve initial surprise on almost each occasion remains a lesson that must simultaneously be reinforced.

Pakistan also overestimated the religious fervour that bound them with the people of J&K; and to whatever extent that it may have been prevalent, this sentiment was crushed under the boots of looting and plundering tribesmen at Baramulla.[7] The incidents that followed not only caused deep anger and disgust in J&K but also amongst the international community at large. As a result, the only realistic possibility of seizing Srinagar was squandered despite an opportunity that was very much present when the tribal foray into J&K began on 22 October 1947.

*1965 Indo-Pak War*

The strategy adopted by Pakistan in 1947–1948 was repeated yet again in 1965, albeit with some modifications. This was possibly a result of the lessons that had been learnt during the previous war. The positive outcome of preliminary military operations undertaken by Pakistan in the Kutch region gave it the confidence to plan a more audacious gambit against India, yet again with the focus on Kashmir.[8] This partly emanated from what was seen as a 'tepid' response by India to Pakistani incursions in the Kutch region.[9]

Pakistan's objective of wresting the valley received a fillip when Kashmir was marked by internal disturbances, commencing in 1963, with the controversy of the missing hair of Prophet Mohammad, considered a holy relic (Moe Muqaddas).[10] Similar to the previous attempt, Pakistan planned to induct a large number of irregular forces into Kashmir, marshalled, led and controlled by the army, to take advantage of the unrest and facilitate the breakdown of a weak administration with support from the population, which they visualised would be supportive of their intervention. Termed as Operation Gibraltar, the force aimed to hit at the soft underbelly of Kashmir, given that the Indian Army was mostly deployed along the line of control (LoC). It was planned that the success of this operation would be followed up with a salami slice in the area of Akhnoor, which represented a vulnerability from India's perspective. If exploited as planned, it could result in cutting off of the state from the rest of the country in terms of surface movement, thereby isolating it. This would have allowed Pakistan to seize control over substantial areas of the state and bargain from a position of strength during the ensuing negotiations.

Based on the lessons that Pakistan had learnt from its previous experience, it carried out important shifts in its preparedness prior to the commencement of the war. First, the atrocities unleashed against the population by the tribals in the earlier war, which alienated the population, stood in contrast with the civil unrest that was prevalent in Kashmir during the period preceding the 1965 war. This gave Pakistan the impression of a conducive support base for its venture. Second, unlike the relatively disorganised tribal marauders, the forces which were a part of Operation Gibraltar were better led and organised.[11] Third, in contrast to the previous attempt, despite differences amongst planners at the initial stage, the entire hierarchy of Pakistan was eventually on board in their attempt to settle the issue of Kashmir on their terms, despite initial hesitation on part of Ayub Khan.[12] Fourth, Pakistan was in a much stronger military position vis-à-vis India, with a large inflow of modern, state-of-the-art weapons that had been supplied to it as a result of its projection as the bulwark of America's anti-communist front. This facade of an anti-communist ideology created by Pakistan was clearly an attempt to bolster its comparative military posture against India. And they largely succeeded in this attempt, as confirmed by the Western Army Commander of India during the 1965 India–Pak War, Lieutenant General (Lt Gen) Harbakhsh Singh, in his book, *War Despatches*.[13] Finally, India's defeat at the hands of China in 1962 led to a belief that a weakened India would not be able to take on a militarily better equipped and motivated Pakistan.

The execution of the plan was not as effective as it had been envisaged. This was because of a number of factors, ranging from poor appreciation of the situation on which the entire plan was based to weak leadership while executing the same. The hybrid conflict rested on three pillars: unrest in Kashmir; breakthrough by the raiders; and finally, success of the conventional attack against Akhnoor. Evidently, Pakistan overestimated the unrest in Kashmir and the expected uprising never took place. J.N. Dixit elaborates that 'Instead of the people in the Valley rising in support of the guerrillas, and acting against their government and security forces, the local population gave continuous and detailed information about the movement of infiltrators to India's security forces.'[14] The initial information of the raiders, in fact, came through locals, who detected some suspicious movement in the area of Gulmarg sector. This ensured that the first pillar of the hybrid strategy collapsed. The ability of the raiders to penetrate into the sensitive areas of Kashmir remained a non-starter given the timely counter-action undertaken by Indian security forces soon after their initial detection. This blunted the second prong as well. Finally, the conventional thrust could still have made substantial headway if Pakistani military leaders had displayed the will to capitalise on the initial breakthrough and head on towards Akhnoor. The indecision that was instead displayed gave the opportunity to India to take counter-measures, thus saving it a potential disaster in the area.

*Switch in Strategy*

The military stalemate of the 1965 Indo-Pak War and the disastrous defeat of the Pakistanis in 1971 thereafter, which resulted in the creation of Bangladesh, made the option of a major military confrontation untenable for Pakistan. The principle focus on a conventional foray supported by sub-conventional forces comprising of irregulars in an environment of popular unrest underwent a shift thereafter. The most prominent element of the same came to the fore in 1988 when mismanagement of affairs in Kashmir, as a result of a flawed electoral process, gave Pakistan the opportunity that became the basis for its hybrid war during the next 30 years.

Unlike the previous two attempts by Pakistan, which were anchored around its conventional attacks, even as the sub-conventional ones provided support in a bid to achieve desired objectives, the disturbances in Kashmir provided suitable conditions for implementation of a sub-conventional-centric approach.

This shift in strategy was not the first by Pakistan. A similar approach had been honed to perfection in Afghanistan along with the United States (US), against Soviet Union, through the employment of 'mujahideen', as they were termed from 1979 till 1989.[15] Simultaneously, though to a lesser degree, a similar experiment was conducted in the Indian state of Punjab, which was experiencing unrest as a result of a separatist movement amongst a section of the population.

Kashmir, however, was different from Punjab. It provided ideal conditions for employment of a large number of elements of hybrid warfare, which became increasingly sophisticated with the passage of time and especially in the information age. It began in 1988 as a separatist struggle spearheaded by the Jammu and Kashmir Liberation Front (JKLF), with backing in terms of arms, training, guidance, funding and diplomatic support from Pakistan. This movement seeking independence was scuttled by Pakistan to prop its proxy terrorist group, the Hizbul Mujahideen (HM), which instead fought for merger of Kashmir with Pakistan. Predictably, this became the favoured terror outfit, with active support by regular armed forces along the LoC through covering fire during infiltration and keeping the Indian Army engaged in firefights. This was accompanied by a well-orchestrated and deeply embedded subversive movement within Kashmir, aimed at hollowing out state institutions and structures, making them suspect in terms of their loyalty and efficacy.[16]

The subversive movement suffered a setback when the HM was largely neutralised by the Indian security forces and Pakistan was forced to reinforce, and at places replace, them with foreign terrorist groups based on members from Pakistan, Afghanistan and volunteers from other countries. Even as these cadres succeeded in raising the military profile of terror groups in Kashmir, their disdain for local life, honour and resources soon alienated them. This led the local population to turn against them. It further strengthened information flow, leading to a large number of these foreign terrorists being neutralised in military operations.

Pakistan, having perceived the regression of Kashmir conflict into international cold storage, took advantage of existing deterrence and India's preoccupation with a number of internal security challenges. It heightened the scale of military adventurism yet again, in a bid to change status quo on Kashmir, by the occupation of high-altitude pickets on the Indian side of the LoC. Both sides, given the challenges associated with sustaining posts at super-high altitude, would vacate these during the winter months every year. This intrusion into Indian territory, once detected in early May 1999, was labelled a mujahideen occupation as part of their 'freedom fight'.

However, the subterfuge did not hold for long as the capabilities on display, intercepted communications and nature of deployment clearly indicated the holding of defences by regular forces.[17] Pakistan had assumed a muted response to the intrusion. They hoped to interdict the road, dominated by these heights and which served as a lifeline to Siachen Glacier, thereby turning the flanks of India's military deployment. However, yet again, the attempt at forcing a decision through a conventional spearhead of the hybrid strategy failed. India reacted with a degree of severity that Pakistan did not expect.[18] The intrusions were slowly but surely reversed from the vital heights. Simultaneously, India's diplomatic offensive exploited the restraint and maturity of the military response, even as a resolute desire to regain all territory occupied was clearly on display. The leadership was also able to convince major powers of Pakistan's deceit and duplicity at Kargil.[19] This two-pronged approached resulted in the military defeat, followed by a complete withdrawal of Pakistani forces without achieving their strategic objectives.

The aftermath of the Kargil conflict saw an upsurge in violence levels. It also witnessed the gradual increase in the employment of informationalised resources as part of Pakistan's hybrid war. As a result, carefully calibrated strategic communications riding on the information highway became the most important tool that was exploited by Pakistan and its proxies in Kashmir. This was conducted with the complete backing of state apparatus and instruments operating from a number of levels. Diplomatically, Kashmir became a consistent agenda for Pakistani leaders at international forums to raise and highlight. This was most prominent at the United Nations (UN) General Assembly[20] and the Organisation of Islamic Cooperation (OIC). Pakistan was instrumental in getting the Contact Group on Kashmir established in 1994. This group consistently raised the issue of Kashmir evidently on behalf of Pakistan. As an illustration, the meeting in 2016 reinforced the 'principled position of the OIC in fully supporting the people of Jammu and Kashmir in their struggle to achieve their legitimate rights.'[21]

After peaking in the first few years of the new millennium, terrorism ebbed by 2012–2013, only to receive a fillip yet again, when the opportunity provided by low levels of violence for ensuring lasting peace and a timely political settlement was missed. Terrorism, which was barely being able to survive the anger of the people, yet again received a boost when the frustrations of some sections within the population were rekindled through a sophisticated campaign by Pakistan and its proxies in Kashmir. Tools of the information age ensured that misinformation at one end and fast dissemination at the other created a lethal mix of indoctrination. This provided the oxygen that the movement had run out of in the last few years. These tools were employed to steer a strategy that employed mass protests by mobs indoctrinated by a virulent ideology and in many cases, paid to create civil unrest.[22]

A more potent employment of the informationalised battlefield was however witnessed in the cyberspace and, more recently, in the social media.[23] Videos were made professionally to support strategic communication that emanated from both Pakistan and within Kashmir. Special cells were created within the Pakistani establishment to further these objectives.[24] The purported real-time videos that were shared inside Kashmir were actually the creation of a sophisticated effort, rather than an amateur one that should have instead been the case. The multiple videos of Burhan Wani, and subsequently the 'human shield' incident, clearly indicated a degree of expertise that could not have been witnessed in real-time situations at the site of incident. This was confirmed by the Ministry of Home Affairs after the analysis of some of these videos that have been shot with multiple cameras, from more than one angle.[25]

The level of organisation and the attempt to galvanise cyber warriors in Kashmir was also indicative of Pakistan's attempts at enlarging the pool of perpetuators in its cyber campaign. This was reinforced on the basis of publicity material circulated in Pakistan by state authorities in a bid to not only enlist but also provide direction to volunteers through the conduct of workshops for undertaking cyber initiatives.[26]

The hybrid nature of war being waged by Pakistan has also manifested itself in the form of various support mechanisms that have been employed to raise funding for terrorism. While the direct use of state apparatus and especially the Inter-Services Intelligence (ISI) is well documented, Pakistan has also employed more innovative methods for this purpose.[27]

The printing of fake Indian currency notes (FICN) was possibly the most important aspect in Pakistan's attempt to impact India economically as well as from the security perspective.[28] While the induction of FICN and its trade facilitated the raising of funds for supporting terrorism, it was equally important for Pakistan to circulate it in the Indian economy with an aim of affecting it adversely. A large number of counterfeit notes in any economy impacts the faith of not only the domestic but also the international monetary climate in relation to that country.

Trade was also exploited through the over/undervaluation of goods to create an illegal surplus value, which could be funnelled for funding terrorism.[29] This was witnessed in case of trade in goods along the border.[30] This trade mechanism was also abused through the smuggling of drugs in large quantity, yet again in an attempt to destabilise the country.[31]

Evidence also came to light which suggested the employment of funding channels like hawala to create unrest in Kashmir.[32] The funds transferred as a result of this safe and opaque system of value transfer revealed the extent to which it was responsible for fuelling the ongoing unrest in J&K.

Pakistan's sub-conventional forays into Kashmir, and beyond into the Indian hinterland, were pitched at levels that tested India's patience and resolve. This was facilitated by Pakistan's nuclear weapons, as an element of its hybrid war. Pakistan commenced its quest for nuclear weapons almost immediately after the humiliating defeat during the 1971 war with India. The military and political elite in the country were convinced that nuclearisation was the only way of achieving deterrence against a conventionally superior adversary.[33] The overt nuclearisation of the sub-continent in 1998 brought the nuclear dimension into sharp focus. This raised concerns of the international community to any military escalation that could result in an intended or more likely an unintended exchange of nuclear weapons.[34] Even as nuclearisation itself had reinforced conventional deterrence, limiting the possibility of a large-scale conventional conflict, it enhanced Pakistan's insurance policy for employing the sub-conventional option with little to deter it.[35] This was brought into focus repeatedly by Pakistan's leadership, which employed nuclear sabre-rattling as a periodic premium for its security insurance policy. It was witnessed during the Kargil conflict; 2001 deployment of the armed forces after the attack on India's Parliament; and in the aftermath of the 2008 Mumbai attack.[36] This provided effectiveness to Pakistan's hybrid war against India at a number of levels. On one hand, it limited India's conventional options against Pakistan and on the other, it gave Pakistan the relatively unrestrained sub-conventional canvas to promote terrorism in Kashmir. Even beyond Kashmir, the aftermath of nuclear tests in Pakistan witnessed sensational terror strikes in India: for example, in the Parliament and Red Fort in Delhi; and in Raghunath Temple and Kaluchak military cantonment, aimed at families of army personnel, in Jammu.

The emphasis on sub-conventional and non-conventional actions as part of hybrid warfare could give an impression that the post-Kargil environment has practically eliminated the conventional element from Pakistan's armoury. This is not entirely true. The post-Kargil phase certainly reduced the intensity of conventional exchanges, but only when compared to the conflict itself. Other than that, firefights between the two armies across the LoC continued unabated. A ceasefire in 2003 did bring down the intensity and frequency of the same for a few years; however, this has been violated all too often to qualify as a robust ceasefire mechanism. An assessment of the ceasefire violation figures given by the government in the recent past clearly reinforces this trend. In 2016, there were a total of 228 ceasefire violations by Pakistan across the LoC and 221 across the international border sector.[37] This is a clear indication of the employment of conventional forces deployed across both the disputed LoC and international border. These incidents cannot be seen in isolation, and given the patterns related to political events, diplomatic exchanges and military incidents, they fit into a larger strategic canvas that is being dominated by the Pakistan Army within the country.[38]

The historical context as well as the more recent approach taken by Pakistan reflects its obsession with India. This has been explained in many ways. From merely a quest for a resolution to the Kashmir issue, to a deeper desire for parity and perhaps superiority over India are some of the others, which seem to bear closer reflection of reality. However, irrespective of the reasons, the methodology repeatedly highlights the employment of hybrid warfare in all its manifestations. These have evolved over the years and become more sophisticated and complex in their application, especially as the last two decades of the information age seem to suggest.

## Pakistan's Hybrid War in Afghanistan

The contours of Pakistan's hybrid war in Afghanistan were crafted under different conditions and against a very different adversary. In fact, the nature of adversary that confronted Pakistan through the course of its campaigns varied from a superior conventional force to an inferior one. It also faced irregular forces during periods in between. On its part, the thrust of Pakistan's foray was led by irregular forces, which were guided and supported by state structures. This displays an inherent flexibility in the approach that was employed. This approach was honed over the years and became a test case for employment elsewhere. Ironically, irregular forces both within and beyond Pakistan have also taken a leaf out of their book to perpetuate a similar kind of conflict upon the masters of the great game.

The first opportunity for Pakistan to employ hybrid warfare in Afghanistan came with the Soviet occupation of the country in 1979. The US was determined to contest the change in status quo at the peak of the

Cold War. Pakistan emerged as a willing and keen partner. It trained and employed the mujahideen at the behest of the US, with funding also coming from Saudi Arabia.[39]

This proved to be an instructive experience for the ISI, which became instrumental in coordinating and implementing the effort. The endeavour provided invaluable lessons related to managing a disparate group of tribals with fluctuating loyalties and an inherent seed of internecine ethnic conflicts. This campaign witnessed religious indoctrination and tribal loyalties as the foundation for mercenaries to be sent into battle. Ahmed Rashid calls it a 'tribal jihad led by clan chiefs and ulema (senior religious scholars) rather than ideological jihad by the Islamicists.'[40]

Pakistan had whetted its appetite, having learnt the difficult art of managing tribal fissures to their advantage. Their biggest success was the victory of their trainees, who, despite having an independent streak, were heavily dependent on Pakistan for the eventual and often spectacular successes on the battlefield in Afghanistan.

> ...the Taliban's closest links were with Pakistan where many of them had grown up and studied in madrassas run by the mercurial Maulana Fazlur Rehman and his Jamaat-e-Ulema Islam (JUI), a fundamentalist party which had considerable support amongst the Pashtuns in Baluchistan and the North West Frontier Province (NWFP). More importantly Maulana Rehman was now a political ally of Prime Minister Benazir Bhutto and he had access to the government, the army and the ISI to whom he described this newly emerging force.[41]

Amongst Pakistan's motivations to extend their influence was the desire to open and control the lucrative route to Central Asia. This route passed through Peshawar, Kabul, Mazar-e-Sharif and Tashkent in Uzbekistan. However, ongoing fighting had blocked it. This led to the option of using the route through Quetta, Kandahar, Herat and Ashkhabad in Turkmenistan. Pakistan's objective was to bribe the warlords controlling this route to commence its usage.[42] The intention was to cut deals through money and subversion, given the limited control that Kabul had over the area.

The military muscle for this venture came in the form of Taliban fighters educated in Pakistani madrasas. On 12 October 1994, approximately 200 fighters moved to the Afghan border to begin their military blitzkrieg into the country.[43] Commencing with a raid on an arms dump under the control of Hekmatyar, the Taliban captured 18,000 Kalashnikovs, artillery pieces, ammunition and vehicles. The small force had heralded its arrival on the scene of action as the primary proxy warriors of Pakistan.[44]

Pakistan's close monitoring of affairs continued through the Kandahar shura—which was attended by their Ambassador at Kabul, Qazi Humayun, and a number of ISI officers—which saw the anointment of Mullah Omar as the *Amir-ul Momineen* or 'Commander of the Faithful'. A similar conciliatory effort was witnessed between 7–13 February 1996, in a bid to bring together all anti-Kabul forces on the side of the Taliban. This was subsequently supplemented with the supply of arms, ammunition, telephone and wireless network and refurbishment of the Kandahar airport.[45] Pakistani madrasas were instrumental in pushing reinforcements at critical stages of the conflict by closing down the religious seminaries for specific durations, as witnessed in August, September and October 1996. The direct involvement of the ISI also included the deputation of officers like Colonel 'Imam' Sultan Amir, who functioned as a district advisor to regional Taliban governments.[46]

Pakistan provided political credibility to the Taliban by recognising its government in early 1997. It also moved its diplomatic and ISI cadres to help them negotiate a settlement in Mazar-e-Sharif after its capture. A large financial package to the tune of $5 million was also provided by Pakistan for the Taliban's campaign in the Bamiyan area, along with substantial support from Saudi Arabia.[47]

Pakistan's gambit not only facilitated its control over Afghanistan though the Taliban, but also provided it with both a breeding and training ground for forces that fought within the country and beyond in Kashmir. This became public knowledge when the US bombed suspected sanctuaries of Osama bin Laden after strikes at their embassies at Kenya and Tanzania in August 1998. The US fired 75 cruise missiles with an aim of hitting Osama at Zawar Kili, Khost. Instead, they hit camps with Pakistanis and Afghans, primarily of the Harkat-ul-Ansar, training to fight in Kashmir.[48] It was also well established that a terrorist camp was located on the outskirts of Kabul at Rish Khor for the Harkat-ul-Mujahideen.[49] The experience in Afghanistan facilitated both the training and blooding in of the terrorists.

While Pakistan was reasonably successful in its military quest in Afghanistan, its actions eventually isolated its progeny, the Taliban, as also their masters, Pakistan. The year 1998 saw the toughest UN sanctions as a result of growing concerns in the US. This was merely the beginning of US pressures on Pakistan and the Taliban, as 9/11 brought the might of the US military machine to their doorstep. Having been given the choice of 'with us or against us', Musharraf, who had by now overthrown Nawaz Sharif in a military coup after the Kargil conflict in May–July 1999, did not blink an eyelid before switching loyalties. From being the first country to recognise the Taliban government, Pakistan was now willing to act as the 'front-line state' in the global war against terror.

One last attempt was made in September 2001 by Lt Gen Mahmud Ahmed, the ISI Chief and the chief interlocutor to broker peace with the Taliban. However, Ahmed's meeting with Mullah Omar failed to convince him of the consequences of refusing to hand over Osama bin Laden to the US. Riaz Mohammad Khan opines that:

> In dealing with the Mujahedin and then the Taliban, the Pakistanis' empathy with their clients made them more than willing to get converted to their clients' point of view than the other way round. Intellectually weak, the midlevel officials, especially those from the ISI, were often impressed and overawed by the certitude of conviction and faith the Taliban demonstrated.[50]

The US-directed and Northern Alliance-led military victory against the Taliban was a severe setback for Pakistan. It lost the enormous leverage that it had enjoyed under the previous regime. In addition, the US forces-led pressure pushed the Taliban and Al Qaeda towards the Afghan–Pakistan border, with a large number seeking refuge in Pakistan's restive tribal areas. Pakistan was caught in a vortex of terrorism and was obliged to take action against these terrorists. The latent domestic terrorists too emerged from their dormant state and the adverse impact of terrorism became enmeshed deeper within the Pakistani society. The delicate balance had been disrupted and the dexterous handling of hybrid warfare had gone out of control.

The diversion of the US attention from Afghanistan allowed Pakistan to support regrouped forces that could further its interests in the country. While the Afghan Taliban and Al Qaeda represented a serious threat for the Afghans and the US-led coalition, for Pakistan, it was the Pakistani Taliban and Al Qaeda that were the targets. This gave the Afghan Taliban a clear field to further its military consolidation in the shadow of Pakistani activities.[51] Under these dichotomous circumstances, Pakistan's resilience and understanding of Afghan politics, as well as the diversion of the US attention towards Iraq, created the perfect conditions for a comeback. This comeback thus stemmed from the US taking its eye off the ball in Afghanistan as a result of their involvement in Iraq to begin with and subsequently, with the public announcement and decision of a drawdown.

The Taliban and Pakistan were clear that the receding US control over the country would offer opportunities that could facilitate the expansion of Taliban influence yet again. This was premised on the understanding of Pakistani Generals, who:

> bury their heads in the sand and pretend that no such threat will materialize (degradation of Taliban and Haqqani network) and that the brinksmanship they pursue with the Americans can pursue indefinitely. Their position is that if they cannot get what they want out of an Afghan settlement, nobody will get an Afghan peace. Blinded by ideology, they resist any forward-looking strategic thinking.[52]

There were clear examples of complicity of the Pakistan Army, and more specifically the ISI, in exploiting the sub-conventional route to retain leverage with the US. This was achieved through the employment of its proxy in the form of terrorist groups like the Haqqani Network. On 10 September 2011, an explosive package sent into a US post killed five Afghans and injured 77 Americans. A few days later, a group of suicide attackers of the Haqqani Network targeted the US Embassy, killing 27 Afghans, thereby humiliating the Americans. This led Leon Panetta and Mike Mullen to indict the ISI and the Haqqani Network. Ahmed Rashid quotes:

> With ISI support, said Mullen, 'Haqqani operatives planned and conducted that truck bomb attack, as well as the assault on our embassy...the Haqqani network acts as a veritable arm of Pakistan's ISI.' Pakistan, Mullen continued, 'may believe that by using these proxies, they are hedging their bets or redressing what they feel is an imbalance in regional power, but in reality, they have already lost that bet.'[53]

Pakistan's ventures in Afghanistan were funded through different sources. As outlined earlier, Pakistan directly funded the Taliban during its ascent within the country, a fact verified by independent sources.[54] However, as the Taliban gained control over Afghanistan, they were able to meet most of their income from local sources. This included extortion and taxation. Ironically, the extortions were from projects implemented in the country that were funded by countries fighting the Taliban. The Taliban also raised a substantial percentage of their total income of approximately $100 million per year from poppy cultivation and trade.[55]

The Haqqani Network's funding pattern in terms of state support was similar. A declassified US Department of State cable suggests that the ISI paid $200,000 to the Network for carrying out the sensational terrorist strike that killed seven Central Intelligence Agency (CIA) operatives in Afghanistan in 2009.[56] Incidentally, the same Network was responsible for the attack on the Indian Embassy in Kabul in 2008, that led to the killing of 58 people. The linkages of the Network do not only emerge with the Pakistani Army but also with groups like the Al Qaeda. The Haqqanis facilitated their escape from Afghanistan after the attacks on Tora Bora.[57]

The historical instances quoted suggest that Pakistan's strategy as part of its hybrid war in Afghanistan hinges on employing proxies like the Haqqani Network and the Taliban. They pursue the policy of subverting sections of the population along the tribal belt of Afghanistan–Pakistan through intolerant, fanatical and rabid misinterpretation of the Islamic faith in order to create a group of brainwashed fighters. This has been funded directly by the Pakistani establishment, as well as criminal enterprises like the Taliban-controlled drug network. Pakistan has also undermined the administration by selectively supporting sections of the tribal population through religious appeals, incentives, bribes and the lure of power.

## Analysis of Pakistan's Hybrid Wars

Pakistan's employment of various constituents of hybrid war remains distinct to each adversary. Both in the case of India and Afghanistan, the constituents of hybrid war have been calibrated keeping in view Pakistan's larger strategic objectives and shifting realities. An assessment of the timeline in relation to both countries suggests that Pakistan had the ability to run two very different campaigns under the aegis of the same institution, its army, in divergent ways, and yet maintain coherence of action. It suggests the evolution of a mature strategy for undertaking hybrid wars by Pakistan, that has stood the test of time, against a lesser military power like Afghanistan and a country like India, that has superior conventional military capability.

The employment of hybrid war has given Pakistan certain distinct advantages that guide its actions in this regard. First, it remains a low-cost option that can be undertaken by an economically weak state, despite a country like India having a distinct advantage in this regard. Second, even as the focus of Pakistan's efforts remains on Kashmir, it allows it to simultaneously affect India's rise as a major power and Afghanistan's very stability as a viable nation-state. This leverage provides Pakistan the ability to influence the environment in its neighbourhood far beyond its real capability. Third, the continuation of war by other means, which is repeatedly kept in public consciousness in Pakistan, allows the army to remain not only relevant but also the sole institution that continues to fight the enemies of the state. This further ensures funds for modernisation and pre-eminence of its leadership. Fourth, hybrid war best allows the exploitation of technological tools that, in a conventional domain, can at best remain pieces of equipment that are placed on display during the annual military parade. However, in the real world of cyberwarfare, information dissemination and distortion allows its actual application on a daily basis. Fifth, given the employment of such tools and proxies, hybrid war also gives a degree of deniability that is either glossed over by supporting powers, pursued inadequately in the absence of obvious state involvement or not considered important enough in light of the low threshold that it functions under.

The most visible and stark constituent of hybrid war is the military arm of a country, especially when applied in a conventional sphere, in a state-on-state conflict. The history of Pakistan's hybrid wars suggests that it has not hesitated to employ this constituent during the past seven decades, when its leadership felt that immediate military and political results could be forced through a favourable position on the battlefield. Despite repeated setbacks, conventional state-on-state use of military force remained the critical element of Pakistan's hybrid war against India in 1947–1948, 1965 and during the Kargil conflict of 1999. However, having failed to achieve the desired results, Pakistan was forced to recalibrate its strategy. With the exception of the Kargil conflict, which was meant to give a fillip to its quest for achieving a more favourable end state in Kashmir, conventional force was applied selectively along the LoC. It became an adjunct and an extension of the sub-conventional application of force. However, even in this form, conventional force has always meant to function as a catalyst, which could enhance the impact of sub-conventional constituents like terrorism. The process of facilitating infiltration into Kashmir has been linked directly with the employment of conventional firepower of the Pakistani regulars on the LoC. Similarly, the strategy of keeping the LoC active through actions by Border Action Teams (BATs) also reflects the exploitation of the army's conventional capability. The use of conventional force has not been as pronounced a component of hybrid war in Afghanistan. While the Pakistani Army has been involved in limited border skirmishes along the Durand Line, the nature and intensity of the same have been lower as compared to the LoC. However, the distinct advantage enjoyed by Pakistan has facilitated the movement of Taliban fighters to and from Afghanistan, thereby assisting their overall operations.

The recent years, as the description of hybrid wars earlier in the chapter suggests, have witnessed an accentuated employment of sub-conventional options by Pakistan. The most prominent amongst these is the use of terrorism as an instrument of state policy. Unlike the past, when sub-conventional means were employed as an auxiliary instrument to its conventional effort, as seen both during the 1947–1948 and 1965 wars with India, the bulk of the heavy lifting has been done by terrorist groups as part of Pakistan's hybrid war. Afghanistan has witnessed the employment of conventional forces in a limited way along the Durand Line and for guidance, training and organisational roles while supporting the Taliban, even as the bulk of the fighting was carried out by guerrilla forces. This was also a result of the nature of adversary in Afghanistan, which did not provide for a clear centre of gravity. This allowed better employment of sub-conventional means rather than conventional, as was witnessed during the early years against India.

Pakistan's direct involvement as part of its endeavour to employ terrorism against India indicates a more intimate involvement in all facets that are linked with the process. The ISI has also been involved in the training of terrorists in camps functioning under their direct supervision. The involvement of the ISI in other support activities, like coordination of highly specialised attacks such as the one witnessed in Mumbai in 2008 and pushing drugs and fake currency, reinforces their salience to the complete cycle of sub-conventional endeavours inside India as an integral part of their hybrid strategic plan. In case of India, Pakistan has wielded greater control over terrorist groups operating in Kashmir, as compared with the Taliban, which has often displayed an independent streak in decision making. However, in contrast with this, the ISI has been involved directly with the Taliban right through the course of conflicts within Afghanistan. This has included guiding their reconciliation efforts, governance, military reinforcements and diplomatic support. Pakistan's recognition of the Taliban government has facilitated their direct involvement further, which distinguishes their role in Afghanistan from Kashmir.

One element of sub-conventional operations that has gained significance in case of India, primarily in Kashmir, is the employment of violent protests. These, as ongoing investigations suggest, have been funded and fuelled by Pakistan in an endeavour to broad base the struggle and give it an indigenous character. It has also facilitated in the creation of a cycle of violence, which generated resentment against the state and simultaneously encouraged recruitment into terrorist ranks. This has remained peculiar to India and has not been evident in the case of Afghanistan.

Pakistan's protracted sub-conventional endeavours received a fillip after the covert and thereafter overt nuclearisation of the country. It was assessed that this would deter India from employing conventional means to punish Pakistan for employing terrorism as state policy. This strategy was further reinforced by Pakistan's refusal to endorse a no-first-use policy, which lowered the nuclear threshold. The deterrence that this ensured gave Pakistan the liberty to operate with impunity, as has been witnessed since the commencement of their sub-conventional endeavour. Nuclear weapons and the threat to employ them as a result of certain red lines being crossed have strengthened Pakistan's hybrid strategy. Statements obliquely threatening the use of nuclear weapons from leaders within the government, during times of confrontation with India, give a sense of rationality emanating from an orchestrated environment of irrationality. The same is not the case in Afghanistan, where Pakistan has not needed the security of its nuclear weapons to pursue its strategic objectives.

Pakistan has also attempted to exploit the economic dimensions of hybrid war. This has been applied in diverse ways against India and Afghanistan. The most obvious has been the state funding of terrorism, for which adequate evidence exists, as has been illustrated earlier in the chapter. Whenever this has not been resorted to, the state has chosen to turn a blind eye to open public fund collection campaigns by groups like the Lashkar-e-Taiba (LeT), Jamaat-ud-Dawa (JuD) and the Falah-e-Insaniat Foundation (FIF). This is despite the fact that all these groups have been listed as terror entities by the United Nations Security Council (UNSC), as well as countries like the US. Pakistan has attempted to make a more serious dent on India's economy through its pumping in of FICN, which is done with the aim of not only funding terrorism but, more importantly, affecting the standing of India's economy. Afghanistan has also faced the brunt of Pakistan's funding and fuelling efforts aimed at supporting terrorism in the country. This has further been aggravated through the drug trade facilitated and supported by Pakistan.

Diplomacy can and has played an important role in Pakistan's hybrid endeavours. The recognition of the Taliban government in 1997 in Afghanistan and eliciting support from countries like Saudi Arabia are some examples of the same. The diplomatic strategy against India is far more elaborate and deep-rooted. There has been a constant attempt by Pakistan to highlight the Kashmir issue at various international forums, as brought out earlier. A series of attempts have been made to convert the bilateral dispute into a UN mandated issue or under the aegis of countries like the US. Further, an attempt has been made to obfuscate the issue of Kashmir with the ongoing policy of state terrorism undertaken inside India by Pakistan's state agencies.

Pakistan's hybrid war against India is primarily against a militarily superior adversary. The repeated inability to take on India on the battlefield has led to the enlargement of the spectrum of conflict by Pakistan. Even though they could not achieve their intended objective of wresting Kashmir or dismembering India, the Pakistani Army has achieved the limited aim of keeping the Indian state tied down in J&K. In the case of Afghanistan, Pakistan is pitched against a weaker adversary. However, despite that, they have found the utility of hybrid warfare in this context as well. It has led them to become the most important external influence in the country.[58] It is difficult to imagine a permanent solution to the instability in Afghanistan without the direct or indirect influence of Pakistan.

The employment of hybrid warfare as relevant to the information age tends to make this phase of conflict different from the past. The manifestation of this shift emerges best from Pakistan's endeavours against India, rather than Afghanistan. The last couple of decades have witnessed a sophisticated employment of instruments of information dissemination and distribution by Pakistan in Kashmir, which cannot be compared to the relatively primitive information domain of Afghanistan. Instead, the latter continues to be impacted by physical methods of individual and group subversion in places of religious dissemination, like the madrasas operating inside Pakistan. The tools employed in the Indian context are not new; however, their scope and effectiveness in an information-fed society has created interconnected players, who can act, react and marshal elements of the conflict faster than ever before. This, as witnessed in Kashmir, can potentially facilitate the creation of conditions that are ideal for mass mobilisation, contrary to the conditions prevalent in Afghanistan. Therefore, even as the hybrid war in Afghanistan is largely a continuing employment of tools from the past, Kashmir has seen its proliferation and evolution. This has witnessed greater reliance on non-kinetic means from kinetic, as was seen in the past. Technology has clearly become the most important driver in this regard, providing tools which can achieve greater effectiveness without the direct use of violence. Ironically, in Kashmir, the instruments of development of the information age have unfortunately become the tools for unleashing destruction within the society by the very people who would have benefited most from peace and stability.

### Table 3.1: Components of Hybrid Warfare Employed by Pakistan

| Component | 1947–48 | 1965 | Kargil | Kashmir | Afghanistan | Remarks |
|---|---|---|---|---|---|---|
| Conventional Force | Yes | Yes | Yes | Yes (limited) | Yes (limited) | Least in Afghanistan |
| Terrorism | No | No | No | Yes | Yes | |
| Armed Intruders | Yes | Yes | Yes | Yes | Yes | |
| Subversion | Yes | Yes | Yes | Yes (social media, Internet, religious indoctrination) | Yes (limited to religious and tribal indoctrination) | |
| Economic Elements | No | No | No | Yes (FICN % direct funds) | Yes (narcotic, direct and external) | From countries like Saudi Arabia |
| Nuclear Threat | No | No | Yes | Yes | No | |
| Cyber | No | No | ? | Yes | ? | |
| Diplomacy | Yes | Yes | Yes | Yes | Yes | |

*Source*: Summary derived from the chapter.

The employment of various components of hybrid warfare can best be encapsulated through a tabular illustration of the same (see Table 3.1).

## Conclusion

Just like the nature of war remains consistent, even as its character changes, hybrid war displays both similarities and differences in case of Kashmir and Afghanistan. The change in its character is more pronounced in Kashmir, given the variety of components employed and the conducive environment that facilitated the same. Afghanistan, in contrast, witnessed changes more in terms of the competing forces that tend to remain in perpetual conflict, even as the nature and character repeatedly carries echoes from the past.

The two case studies suggest that successful use of hybrid war is dependent on employing the right tools in an appropriate context. Pakistan's initial insistence on using conventional means against India did not yield results commensurate to the resources applied. Conversely, their choice of relying more on sub-conventional means both in Afghanistan and India gave them greater success, at least in the short term, without co-relating it with the impact that nurturing terrorism has had on their own society.

Hybrid war, when employed through reliance on sub-conventional means, also tends to protract conflicts, as seen both in Afghanistan and

India. Given the changing character of war, this could well indicate the future course of similar conflicts witnessed in West Asia as well.

When viewed from the perspective of the target country, in this case both Afghanistan and India, the response to Pakistan's hybrid war has been varied. While it is beyond the scope of this chapter to elaborate upon the same, it does need emphasis that this too would go beyond the traditional tools employed to defend a country against conventional military threats. The all-of-government response that it demands would be indicative of the quality and effectiveness of the same. This is apparent in both cases. Afghanistan, given its weak state structures, has not been able to handle Pakistan's hybrid war as well as India, which has not only the resources but also the comprehensive national resilience to blunt and perhaps even take the fight back to the perpetuator.

## NOTES

1. C. Christine Fair, *Fighting to the End: The Pakistan Army's Way of War*, New Delhi: Oxford University Press, 2014, p. 6.
2. Josef Korbel, *Danger in Kashmir*, Karachi: Oxford University Press, 2002, p. 121.
3. Akbar Khan, *Raiders in Kashmir*, Delhi: Army Publishers, 1990, pp. 11–32.
4. Agha Humayun Amin, *The 1947–48 Kashmir War: The War of Lost Opportunities*, March 1999, available at https://archive.org/stream/The1947-48KashmirWar TheWarOfLostOpportunities/49202996-The-1947-48-Kashmir-War-Revised_djvu.txt, accessed on 6 June 2017.
5. Ibid.
6. In contrast with the thought expressed by Akbar Khan, British policy was clearly in favour of Pakistan, which was considered a more pliable and useful ally to further British interests in the Middle East and South Asia. This was echoed repeatedly prior to and during the 1947–1948 operations. See C. Dasgupta, *War and Diplomacy in Kashmir: 1947–48*, New Delhi: Sage, 2002, pp. 14–19, 57–62.
7. There were exceptions to this in areas like Poonch, where local dissent amongst tribesmen, who were up in arms due to agrarian issues, was exploited.
8. J.N. Dixit, *India–Pakistan in War & Peace*, New Delhi: Books Today, 2002, p. 144.
9. Fair, *Fighting to the End*, n. 1, p. 141.
10. Altaf Gauhar, *Ayub Khan: Pakistan's First Military Ruler*, Dhaka: Dhaka University Press, 1996, p. 204.
11. Ibid., pp. 213–214.
12. James P. Sterba, 'Bhutto Picks up the Pieces of Pakistan', *The New York Times*, 25 June 1972, available at http://www.nytimes.com/1972/06/25/archives/bhutto-picks-up-the-pieces-of-pakistan-bhutto-picks-up-the-pieces.html, accessed on 10 August 2017.
13. Harbakhsh Singh, *War Despatches: Indo-Pak Conflict 1965*, New Delhi: Lancer International, 1991, p. 7.
14. Dixit, *India–Pakistan in War & Peace*, n. 8, p. 146.
15. See Levy and Scott, The Dangerous Relationship between United States and Pakistan, New York: Walker Books, 2008, p. 182.

16. See Jagmohan, *My Frozen Turbulence in Kashmir*, New Delhi: Allied Publishers Pvt. Ltd, 1996, p. 375.
17. See Jaswant Singh, *A Call to Honour: In Service of Emergent India*, New Delhi: Rupa & Co., 2006, pp. 212–219.
18. Sartaj Aziz, *Between Dreams and Realities: Some Milestones in Pakistan's History*, Karachi: Oxford University Press, 2009, p. 259. The author quotes Musharraf, who acknowledged that he did not expect the nature of retaliation that came from India.
19. The US as well as the G-8 countries clearly conveyed, in their respective statements, their displeasure with any attempt to change status quo at Kargil. See 'The Kargil Conflict', *Pakdef*, available at http://pakdef.org/the-kargil-conflict/, accessed on 19 July 2017.
20. See General Assembly of the UN, Speech of Prime Minister Nawaz Sharif, 26 September 2016, available at http://www.un.org/en/ga/69/meetings/gadebate/26sep/pakistan.shtml, accessed on 10 June 2017.
21. 'Report of the Meeting of the Contact Group on Jammu and Kashmir', New York, 19 September 2016, available at http://www.oic-oci.org/upload/conferences/acm/2016/en/JK-REP-FINAL-ACM2016-EN.pdf, accessed on 10 June 2017.
22. Jamshed Khan and Sushant Pathak, '"This is Our Bread and Butter": Undercover Reporters Film Kashmiri "Stone-pelters" Admit to being Paid Rs 7,000 a Month for Throwing Stones and Molotovs at Indian Forces', *Mail Online India*, 29 March 2017, available at http://www.dailymail.co.uk/indiahome/indianews/article-4362020/Kashmiri-stone-pelters-admit-paid-film.html, accessed on 19 July 2017.
23. For an assessment of Pakistan's cyber campaign in the aftermath of the surgical strikes, see Aaditya Purani, 'How India–Pakistan Hackers Escalated Cyber War Post Surgical Strikes', *DailyO*, 12 October 2016, available at http://www.dailyo.in/politics/india-pakistan-war-cyber-security-national-green-tribunal-hackers/story/1/13367.html, accessed on 10 June 2017.
24. Amit Khajuria, 'Pakistan Intensifies Cyber Warfare over Kashmir', *The Tribune*, 22 April 2017, available at http://www.tribuneindia.com/news/jammu-kashmir/community/pakistan-intensifies-cyber-warfare-over-kashmir/395539.html, accessed on 10 June 2017.
25. Ibid.
26. Kartikeya Sharma, 'DNA Exclusive: Pakistan Organising Cyber Workshops to Instigate Unrest in Kashmir', *DNA*, 1 April 2017, available at http://www.dnaindia.com/india/report-cyber-war-is-the-new-face-of-insurgency-in-kashmir-2378034, accessed on 10 June 2017.
27. For a detailed assessment of Pakistan's funding of terrorism in India, see Vivek Chadha, *Lifeblood of Terrorism: Countering Terrorism Finance*, New Delhi: Bloomsbury, 2015.
28. See Narayan Lakshman, 'Fake Currency from Pakistan Threat to Indian Economy: U.S.', *The Hindu*, 5 March 2011, available at http://www.thehindu.com/news/international/Fake-currency-from-Pakistan-threat-to-Indian-economy-U.S./article14935520.ece, accessed on 14 June 2017.
29. For details on trade-based money laundering and terrorism, see 'APG Typology Report on Trade Based Money Laundering', Asia/Pacific Group on Money Laundering, 20 July 2012, available at http://www.fatf-gafi.org/media/fatf/documents/reports/Trade_Based_ML_APGReport.pdf, accessed on 14 June 2017.

30. 'NIA Probes Almond Trade at LoC for Possible Terror Funding', *Business Today*, 19 January 2017, available at http://www.businesstoday.in/current/economy-politics/nia-probes-almond-trade-at-loc-examining-voluminous-documents/story/244534.html, accessed on 14 June 2017.
31. 'Drugs Seized from PoK Truck in Uri', *The Tribune*, 7 February 2015, available at http://www.tribuneindia.com/news/jammu-kashmir/crime/drugs-seized-from-pok-truck-in-uri/38887.html, accessed on 14 June 2017.
32. Rajesh Ahuja and Toufiq Rashid, 'NIA Finds Rs 2.5 crore, LeT Letterheads in Raids on Separatists, Hawala Dealers', *Hindustan Times*, 4 June 2017, available at http://www.hindustantimes.com/india-news/nia-raids-14-places-in-kashmir-8-in-delhi-over-terror-funding-from-pakistan-files-fir-against-lashkar-founder-hafiz-saeed-separatist-geelani/story-flUrFleQ2esT2RY10mhDKK.html, accessed on 14 June 2017.
33. Feroze Hassan Khan, *Eating Grass: The Making of the Pakistani Bomb*, New Delhi: Foundation Books, 2013, p. 126.
34. Stephen Cohen, 'Nuclear Weapons and Nuclear War in South Asia: An Unknowable Future', The Brookings Institution, 1 May 2002, available at https://www.brookings.edu/wp-content/uploads/2016/06/cohens20020501.pdf, accessed on 20 July 2017.
35. Owen L. Sirrs, *Pakistan's Inter-Services Intelligence Directorate: Covert Action and Internal Operations*, New York: Routledge, 2016, p. 166.
36. See Dixit, *India–Pakistan in War & Peace*, n. 8, p. 61; Cohen, 'Nuclear Weapons and Nuclear War in South Asia', n. 35.
37. Lok Sabha Unstarred Question No. 2638, 'Ceasefire Violations by Pakistan, answered on 17 March 2017, available at http://164.100.47.190/loksabhaquestions/annex/11/AU2638.pdf, accessed on 14 June 2017. Initial data for the year 2017 suggests a more than two-fold increase in these numbers.
38. This phenomenon has often been explained through the analogy of the army having a nation rather than the reverse, which is otherwise true of most democratic countries. Christine Fair develops this theme in her book, *Fighting to the End*, n. 1, pp. 27–31.
39. For a first-person account of Pakistan's role in Afghanistan during this phase, see Mohammad Yousaf and Mark Adkin, *Afghanistan: The Bear Trap: The Defeat of a Superpower*, New Delhi: Variety Book Depot, 2006.
40. Ahmed Rashid, *Taliban: Islam, Oil and the New Great Game in Central Asia*, London: I.B. Tauris, 2000, p. 18.
41. Ibid., p. 26.
42. Ibid., pp. 26–27.
43. Ibid., p. 27.
44. Ibid., p. 28.
45. Ibid., pp. 42, 44, 45.
46. Michael Rubin, 'Who is Responsible for the Taliban?', The Washington Institute, March 2002, available at http://www.washingtoninstitute.org/policy-analysis/view/who-is-responsible-for-the-taliban, accessed on 20 July 2017.
47. Rashid, *Taliban: Islam, Oil and the New Great Game in Central Asia*, n. 40, p. 72.
48. Riaz Mohammad Khan, *Afghanistan and Pakistan: Conflict, Extremism and Resistance to Modernity*, New Delhi: Oxford University Press, 2011, p. 81.
49. Rubin, 'Who is Responsible for the Taliban?', n. 46.

50. Khan, *Afghanistan and Pakistan*, n. 48, p. 91.
51. Ibid., p. 142.
52. Ahmed Rashid, *Pakistan on the Brink: The Future of Pakistan, Afghanistan and the West*, London: Penguin, 2013, p. 161.
53. Ibid., pp. 180–181.
54. See 'Pakistani Agents Funding and Training Afghan Taliban', *BBC News*, 13 June 2010, available at http://www.bbc.com/news/10302946, accessed on 20 June 2017.
55. 'Who is Funding the Afghan Taliban? You Don't Want to Know', *Global Post*, 13 August 2009, available at http://blogs.reuters.com/global/2009/08/13/who-is-funding-the-afghan-taliban-you-dont-want-to-know/, accessed on 20 June 2017.
56. Lalit K. Jha, 'ISI Funded Haqqani Network to Attack CIA Camp in Afghanistan', *India Today*, 14 April 2016, available at http://indiatoday.intoday.in/story/isi-funded-haqqani-network-to-attack-cia-camp-in-afghanistan/1/643090.html, accessed on 20 June 2017.
57. Mohammad Taqi, 'Pakistani Patronage of Haqqani Network Continues Undeterred as US Turns Blind Eye', *The Wire*, 18 April 2016, available at https://thewire.in/30099/pakistani-patronage-of-haqqni-network-continues-undeterred-as-us-turns-a-blind-eye/, accessed on 22 June 2017.
58. The Pentagon, in its six-monthly report, identifies India as the most reliable and Pakistan as the most influential external factor in Afghanistan. See Simran Sodhi, 'India Afghanistan's Most Reliable Partner: Pentagon', *The Tribune*, 22 June 2017, available at http://www.tribuneindia.com/news/nation/india-afghanistan-s-most-reliable-partner-pentagon/425756.html, accessed on 10 July 2017.

# 4

# Russia and Hybrid Warfare
## Achieving Strategic Goals without Outright Military Force

*Aman Saberwal*

Russia is the largest country in the world and is approximately six times the size of India. In terms of population, it has the ninth largest population globally, roughly one-tenth that of India. The country has an ageing populace with a negative population growth rate. It has a gross domestic product (GDP) of roughly $1.5 trillion (in purchasing power parity [PPP] terms, it is over $3 trillion).[1] Energy plays a big role in its economy. Low oil prices have affected its economy, but the country is still very much relevant globally. In the view of the United States (US) and its European allies, Russia has been punching way above its weight in international politics after the fall of the Soviet Union. In fact, Russia's latest diplomatic success has been its role and active military involvement in Syria in support of President Assad, much to the chagrin of the West. The details of Russians actions in Syria are not discussed further in this chapter as they have been covered in the next one.

The Russians have historically been past masters of disinformation. The story of the Potemkin villages in the late eighteenth century (whether historical fact or exaggeration)—which were readied overnight and shifted after the Empress, Catherine II, left to give her the impression that all was well—bears this out. This propensity for disinformation reached its pinnacle in the Soviet era.[2] As Lenin said in 1920:

> One will readily agree that any army which does not train to use all the weapons, all the means and methods of warfare that the enemy possesses, or may possess, is behaving in an unwise or even criminal manner. This applies to politics even more than it does to the art of war.[3]

The current Russian thinking can be encapsulated by the so-called Gerasimov Doctrine, which was actually a 2,000-word article by the then Russian Chief of General Staff in the weekly Russian trade paper, *Military–Industrial Kurier*, in February 2013. It articulated the use of all available means to achieve a political goal. A few lines are quoted from the translation of the article to illustrate what he meant to say:

> In the 21st century we have seen a tendency toward blurring the lines between the states of war and peace. Wars are no longer declared and, having begun, proceed according to an unfamiliar template...The very 'rules of war' have changed. The role of non-military means of achieving political and strategic goals has grown, and, in many cases, they have exceeded the power of force of weapons in their effectiveness...All this is supplemented by military means of a concealed character, including carrying out actions of informational conflict and the actions of special-operations forces. The open use of forces—often under the guise of peacekeeping and crisis regulation—is resorted to only at a certain stage, primarily for the achievement of final success in the conflict.[4]

From the given enunciation flows the concept of non-linear warfare, which can be used to try to understand the current Russian military thinking. As per this concept, there is no distinct conflict zone and it does not have clear front lines or distinct friendly/enemy areas. Non-linear warfare relies on the subversion and division of the enemy's social and political structure, allowing the aggressor to do their will by any means, not just brute force. It has no bounds and sometimes functions with limited planning. It employs many measures that would not conventionally seem like warfare, such as propaganda, political and social agitations and cyberattack. This does not mean kinetic action or hard force remains unutilised, but it is supplemented by these measures. The key in using these approaches is making it unclear what is going on. These approaches can be best understood by looking at Russian actions in various theatres that show how these steps forward their strategic objectives, best explained by Russian actions in Crimea, Estonia and Ukraine.[5]

## Crimea

Crimea is a strategically important peninsula on the north coast of the Black Sea in Eastern Europe. Ethnic Russians make up the vast majority of the population of the Crimean Peninsula, with significant Ukrainian and Crimean Tatar minorities. It is known as the setting for the epic poem, 'Charge of the Light Brigade'. It has an interesting and chequered history. In 1783, Crimea became a part of the Russian Empire as a result of the Russo-Turkish War (1768–1774). Following the Russian Revolution of 1917, Crimea became an autonomous republic within the Union of Soviet Socialist Republics (USSR). During the Second World War, Crimea was downgraded to Crimean oblast and then, in 1954, it was transferred to Ukraine.[6] When the Soviet Union collapsed in 1991, Ukraine became an independent country and the Crimean Peninsula became an autonomous republic within Ukraine.

The erstwhile Soviet Black Sea Fleet was headquartered at Sevastopol, the largest city on the peninsula, and the successor Russian Navy Black Sea Fleet also continued to be based there. The Ukrainian naval forces were also co-located there.

The Crimean Peninsula holds a special significance for Russian imagination and self-awareness[7] as Sevastopol saw some of the bloodiest battles in the Second World War between the Russians and the Nazis. Some Russian historians hold that the Crimean Peninsula was transferred to Ukraine by Khrushchev to obtain the support of the Ukrainian Communist Party, the second biggest party and a powerful entity in the then Soviet Union, for the Chairmanship of the Communist Party.[8] Crimea became the epitome of disenchantment and alienation felt by ethnic Russians and it registered the lowest rates of approval in the referendum held on Ukrainian independence in 1991. In independent Ukraine too, there was a lack of feeling of belonging to Ukraine in the minds of the Russian-origin population of Crimea.[9] It was thus not surprising that the Russian moves, post the February 2014 Ukrainian Revolution and ouster of President Viktor Yanukovych, received widespread local support and the controversial referendum received an overwhelming vote for secession from Ukraine.

After Ukraine broke away from the erstwhile Soviet Union and declared independence, various ethnic Russian politicians had sought to assert their sovereignty and strengthen ties with Russia through various steps, which were then declared unconstitutional by the Ukrainian government. The 1996 Constitution of Ukrainian had stipulated that Crimea would have the status of an autonomous republic, but insisted that Crimean legislation be in keeping with that of Ukraine. The February 2014 Ukrainian Revolution, which ousted President Viktor Yanukovych, had also resulted in a political crisis in Crimea.

Demonstrations broke out against the new Ukrainian government. The incumbent Crimean prime minister however maintained, against popular mood, that he supported the new government in Ukraine. Large-scale protests were held throughout Crimea by ethnic Russians against the new government, and also by various supporters of the new government. In Sevastopol, protestors voted to establish a parallel government. They also created civil defence squads with the help of the Russian Night Wolves motorcycle club. The Night Wolves were reportedly funded by the Kremlin and were supporters of President Putin.[10] The Russian president had even been reported to have ridden with them earlier. *Time* magazine reported at that time that the leader of the Night Wolves, Alexander Zaldostanov, who was an old friend of President Putin, had come to Crimea during the crisis period. The Night Wolves were later subjected to US sanctions for their role in Crimea and Ukraine.

The pro-Russian protestors waved Russian flags and called Putin their president. Russian military convoys were also reportedly seen in the area. On 25 February, hundreds of pro-Russian protesters blockaded the Crimean Parliament demanding a referendum on Crimea's independence. Sevastopol elected a Russian citizen as mayor, even as the incumbent appointee by the Ukrainian president continued to hold the post. Large-scale demonstrations continued, both by Ukrainian and Russian supporters, leading to clashes between them. Russian troops secured the main route to Sevastopol reportedly on direct orders from President Putin. A military checkpoint, duly flying the Russian flag, was set up on the main highway between Sevastopol and Simferopol.[11]

On 27 February, Russian incognito special forces seized the buildings of the Crimean Parliament and the Crimean Council of Ministers in its capital, Simferopol.[12] Russian flags were then raised over these buildings. These were the infamous 'little green men', that is, armed uniformed men without insignia. At that time, Russia denied that these were their soldiers; however, later, President Putin accepted that these were Russian military personnel and justified their use.[13] Crimea was then effectively cut off from Ukraine by means of various checkpoints established at the borders. The Parliament held an emergency session and controversially voted to replace incumbent Prime Minister Anatoli Mohyliov with Sergey Aksyonov of the Russian Unity Party and hold a referendum on autonomy.

According to the Constitution of Ukraine, the Prime Minister of Crimea was appointed by the Supreme Council (Parliament) of Crimea in consultation with the President of Ukraine. Both newly appointed Crimean Prime Minister Aksyonov and Speaker Vladimir Konstantinov held that they viewed the deposed Viktor Yanukovych as the *de jure* President of Ukraine. Aksyonov then took over control of all Ukrainian military installations in Crimea. On 1 March, he formally asked Russia for assistance to ensure peace in Crimea. The Kremlin then rushed more Russian forces to Crimea. These, again, were the ubiquitous little green men and operated without insignia. Russia was in firm control of Crimea and it had been cut off from Ukraine. The Russian aim had effectively been achieved.

Ukraine accused Russia of having its forces in Crimea in violation of treaty obligations. However, Russia denied these allegations.[14] The controversial referendum, advanced to 16 March, resulted in 95 per cent of the population of Crimea voting in favour of Russian control over the peninsula.

The Russians exploited divisions within the Crimean society to their advantage and used the Russian ethnic groups and their historical ties to further their strategic aims. They used proxies and groups like the Night Wolves motorcycle club to achieve their aims. They sent in soldiers without insignia, the ubiquitous little green men, which could have been a deniable intervention had they not succeeded, and their use was justified later by legal rhetoric.[15] The cyber domain was also used by the Russians effectively. Cyberattacks were carried out on Ukrainian phone and mobile networks. Landlines and mobile phones were not available throughout Crimea during the period of the crisis.[16] Internet in Crimea was totally controlled by Russia. Russian ships were suspected of having jammed Ukrainian radio communications affecting command and control of their forces.[17]

Russia thus achieved its aims in Crimea in a classic exposition of non-linear warfare, as expounded by Gerasimov.

## Estonia

Estonia is one of the Baltic republics which was incorporated into the Soviet Union in 1940. The Russian ethnic minority forms 26 percent of the population of Estonia. After dissolution of the Soviet Union, Estonia regained its independence and started on the path of economic, political and social reforms. It joined the European Union (EU) and North Atlantic Treaty Organization (NATO), much to the chagrin of Russia. Discussing Estonia in the context of Russian non-linear warfare will help us to understand the evolution of the Russian thought process, as it indicates Russian revival and return of confidence after the events of the break-up of the Soviet Union.

In April 2007, tensions with Russia increased due to the decision of Estonian capital city (Tallinn) authorities to remove the statue of the Bronze Soldier of Tallinn, which commemorated Soviet soldiers who had liberated Estonia during the Second World War. Many Estonians, however, saw it as a symbol of oppression. For Russians, it meant the destruction of their cultural heritage and lack of respect for the Red Army which fought against Nazi Germany during the Second World War. After the removal of the statue, the relationship between Estonia and Russia became strained.

Russia accused the Tallinn authorities of breaking human laws and demanded resignation of the Estonian prime minister. There were riots on the streets involving the police and the Russian minority, protests in front of the Estonian Embassy in Moscow and massive cyberattacks on Estonia. Estonia was almost entirely digitised and highly dependent on the Internet even in 2007. Almost the whole country was covered by Wi-Fi and Internet. All government services were available online. Majority of the Estonian population did their banking transactions online. It was thus highly vulnerable to cyberattacks.

The cyberattacks started on 26 April, peaked on 9 May and thereafter petered off with the last attack on 23 May. The attacks were distributed denial-of-service (DDOS) attacks and were later traced to Russia. The DDOS attacks successfully targeted the websites of all government ministries, two major banks and several political parties. The hackers were even able to disable the parliamentary email server, and also disabled credits card machines and automated teller machines (ATMs) throughout Estonia. The country was brought to its knees without any direct attack through a new type of action. The world took notice of the cyberattacks and the vulnerabilities exploited. The services were restored gradually with the assistance of various countries, including Germany, Israel, Slovenia and Finland. The NATO computer emergency response team (CERT) was also involved and all concerned learned a valuable lesson.

Though a relatively small and isolated incident, this was a stepping stone for Russia as it helped hone its cyber capabilities and doctrines for non-linear warfare and hybrid warfare. This was also a precursor to Georgia and many aspects of cyber and hybrid warfare tried out here were put to good use in Georgia the following year.

## Georgia

Georgia is a Caucasian country located to the south of Russia. It has Turkey to its south and the Black Sea washes its western shores. It was part of the erstwhile Soviet Union and regained its independence after its break-up. After it gained independence, Georgia slowly started gravitating towards the Western sphere of influence. This trend gathered further momentum after the 'Rose Revolution' of 2003 (in which thousands of people hit the streets to protest against the flawed results of a parliamentary election), which saw the overthrow of the incumbent President Eduard Shevardnadze who was an ex-Soviet Foreign Minister.

Abkhazia and South Ossetia are enclaves geographically within Georgia in the Caucasus, claiming independence from Georgia. The two regions recognise each other, and also have been recognised by a few other states, including Russia, which recognised the independence of Abkhazia and South Ossetia after the Russo-Georgian War. Georgia and the majority of countries of the world do not recognise them as independent states. Georgia officially considers them as sovereign territory of the Georgian state under Russian military occupation.

South Ossetia and Abkhazia were Russian-supported breakaway Georgian provinces. Both provinces had the presence of Russian peacekeepers. Georgian President Mikheil Saakashvili's attempts to reintegrate the provinces led to a strong Russian reaction and the war of 2008.

The Georgian Army entered South Ossetia on 7 August 2008 after clashes with South Ossetian separatists. Russia accused Georgia of aggression against South Ossetia and entered the conflict the next day. The Russian Army launched attacks into Georgia, ably supported by the air force and the navy. Russian and South Ossetian forces battled Georgian forces in and around South Ossetia for several days, until the Georgian forces were forced to retreat.

Russian and Abkhaz forces, meanwhile, opened a second front against Georgia in Abkhazia. The Russian naval forces enforced a blockade of the Georgian coast. The Russian Air Force easily achieved air superiority and also attacked targets well beyond the conflict zone, deep into Georgia. A ceasefire was declared on 12 August between the warring parties.

This was the first conflict in which cyberwarfare and military force were used concurrently, and the cyber element augmented and enhanced the effect of military action. Cyberattacks, on a small scale, were being mounted on Georgian sites in the months leading up to the conflict. On 19

July 2008, security firms issued a warning about a DDOS attack on several Georgian websites. A similar attack, on a much bigger scale, was launched on 8 August 2008, and the date coincided with the Russian entry into the conflict.

The cyberattack carried out by the Russian hackers can be divided into two phases. In the first phase, the hackers focused mainly on Georgian news and government websites. They used botnets—a network of private computers infected with malicious software and controlled as a group without the owners' knowledge—to conduct mainly brute DDOS attacks which crippled the target sites. The Georgian networks were more vulnerable to attack than the Estonian ones and the lessons from Estonia were also put to good use. In the second phase of the cyberattacks, the list of targets expanded to include financial institutions, businesses, educational institutions, Western media and even a Georgian hackers website. Beside the DDOS attack, there were also Web defacement operations and massive spamming on public email networks in order to clog them. During the second phase of operations, a lot of 'patriotic hackers' also joined in the campaign against Georgian sites.

Till 10 August, the majority of the Georgian governmental websites were down and the Georgian government was unable to communicate with the world using the Internet. Even the Georgian president's website had been defaced and depicted him as Hitler. Banks and mobile phones throughout the country were paralysed.

The attacks allegedly originated in Russia and were reportedly by both professionals and patriotic Russian hackers. Conventional attacks and cyberattacks coincided and there were no physical attacks on communication and media facilities, leaving them to be taken down by cyberattack alone. These attacks cut off information flow to Georgian residents, leaving them open to propaganda. Thus, these cyberattacks on Georgia brought normal life to a standstill and hastened the end of the war by reducing the war-waging potential of Georgia.

The war with Georgia is another good example of how the Russians effectively prosecute a hybrid war to further their strategic aims. They used the Georgian vulnerabilities to good effect and kept increasing their influence in the enclaves of South Ossetia and Abkhazia. Their proxies, the separatists in South Ossetia, engaged the Georgian forces and this led to the Georgians invading to stop these attacks, thus providing an opportunity to Russia to intervene and prevail militarily over Georgia not only in South Ossetia but also in Abkhazia to establish them as entities independent of Georgia. Russia utilised its cyber capabilities and experience of Estonia to good effect to complement military capabilities. Denying the use of the Internet to communicate its version of events to Georgia enabled the Russians to utilise their capabilities of information warfare and propagate their point of view effectively.

## The Advantages of Hybrid Warfare for Russia

The erstwhile Soviet Union was a superpower and as the successor state, Russia boasted of considerable military capability. This was in stark contrast to Estonia and Georgia and even Ukraine. Russia had overwhelming military superiority against these adversaries. In spite of that, it chose not to exercise the military option at all in Estonia and Ukraine (apart from using its troops as the 'little green men') and used it in a limited way even in Georgia. What did it achieve through this?

The biggest advantage that accrued to Russia was that it could achieve its objectives without major international intervention. Had it openly used military force in any of the above-mentioned situations, it may have had to contend with immediate international condemnation and United Nations (UN) scrutiny, even though as a permanent member of the UN Security Council it could have vetoed any measures majorly prejudicial to its interests. Its actions in Estonia were limited to the cyber domain and sent a message to the former constituents of the Soviet Union that it still wielded considerable influence. The world was also made aware of Russian cyber capabilities.

In Georgia, Russia effectively propped up separatists in South Ossetia and Abkhazia, and even had its forces in the two enclaves for 'peacekeeping'. It had an excuse for attacking Georgia once the separatist firing from South Ossetia took a toll on Georgian forces and they went on the offensive against the separatists. It could thus take the high moral ground in spite of the fact that its proxies in Abkhazia took the opportunity to take on the Georgian forces, and Russian Armed Forces—army, navy and air force—attacked Georgia. In a swift action in which its cyber capabilities effectively choked Georgia from putting its point across, it achieved its aims and the enclaves of South Ossetia and Abkhazia are effectively under its influence now. By using hybrid strategies, it achieved what it had set out to do with minimum outside interference and could also ride out the protests of the international community.

The biggest advantage of using hybrid warfare strategies accrued to Russia in Crimea. The peninsula had a great strategic value and the events in Ukraine, which had been a part of the erstwhile Soviet Union, and the majority ethnic Russian population gave Russia an opportunity which it seized. Its intervention initially was highly deniable as its troops did not wear any insignia (as mentioned earlier) and it used proxies, such as the Night Wolves, which were not directly linked to the Russian state apparatus to further its interests. Thus, if things had not gone its way, the Russians could have claimed that they had not participated in the events there. This degree of deniability is one of the biggest things in favour of hybrid warfare vis-à-vis traditional armed conflicts. The Russians could effectively counter any criticism of their actions in Crimea by just denying involvement. It is pertinent to mention here that the Russian 'Liberation of Crimea' medal was instituted for actions from 20 February to 18 March 2014. That was two days before President Yanukovych fled Kiev, the Ukrainian capital, and was deposed.[18]

Russia's use of information warfare, of which it was the past master, including what is now popularly called 'Lawfare', that is, twisting legal facts to suit its convenience, was also very effective as the deposed Viktor Yanukovych was considered as the *de jure* President of Ukraine and a new Prime Minister of Crimea was appointed, Aksyonov, who was totally pro-Russian and enabled quick and legally tenable Russian control of the peninsula after a pro-Russian referendum. Thus, Russia could turn its *defacto* control of the Crimean Peninsula into one that could stand up to international scrutiny as a lawful action of an independent state whose citizens wanted closer ties with Russia. The various actions outlined earlier in the chapter ensured that Ukraine, which was in the midst of a political upheaval itself, was effectively out manoeuvred in a manner that was not patently against established norms of international conduct, especially as it provided a fig leaf of legality to Russia. Thus, Russia achieved its aims without resorting to a declared war. That was the greatest advantage of using the elements of hybrid warfare, or non-linear warfare in the Russian parlance, for the Russians.

## Conclusion

Russia has been extremely successful in applying the concepts of non-linear war (which is very similar to hybrid warfare referred to by Western sources), as expounded by Gerasimov, in furthering its strategic aims. It has been nimble in changing its means as required to achieve its aims. Russians have mastered the cyber domain and have used it to their advantage to further their interests wherever pertinent. They are very clear about the desired end state and can withstand extreme diplomatic pressure while striving to achieve it. They can be economical with the truth when its suits them, as they displayed in Crimea by denying their involvement and disowning the 'little green men' while the crisis unfolded and accepting that they were their armed forces personnel later.

Their readiness to use proxies as required, and follow up with the conventional force of arms at the appropriate time, epitomises their willingness to adapt to the changing battlefield. Ultimately, doing all that has to be done by combining kinetic and non-kinetic means, military and non-military means, to achieve the goals underlines what hybrid warfare stands for.

## NOTES

1. World Bank data, available at https://data.worldbank.org/country/russian-federation. Accessed 7 February 2018.
2. Ion Mihai Pacepa and Ronald J. Rychlak, *Disinformation: Former Spy Chief Reveals Secret Strategies for Undermining Freedom, Attacking Religion, and Promoting Terrorism*, WND Books, 25th June 2013, USA.
3. *Left-Wing Communism: An Infantile Disorder*, Collected works of Lenin, 4th Edition, Volume 31, p. 96.
4. Available at inmoscowsshadows.wordpress.com/2014/07/06/the-gerasimov-doctrine-and-russian-non-linear-war/. Accessed 7 February 2018.
5. Mark MacKinnon 'Russian-backed Fighters Restrict Access to Crimean City', *The Globe and Mail*, 26 February 2014. https://www.theglobeandmail.com/news/world/tension-in-crimea-as-pro-russia-and-pro-ukraine-groups-stage-competing-rallies/article17110382/. Accessed 7 February 2018.
6. https://www.wilsoncenter.org/publication/why-did-russia-give-away-crimea-sixty-years-ago. Accessed 7 February 2018.
7. Sergey Saluschev, 'Annexation of Crimea: Causes, Analysis and Global Implications', *Global Societies Journal*, Vol. 2, 2014. https://escholarship.org/uc/item/5vb3n9tc. Accessed 7 February 2018.
8. Ibid.
9. Ibid.
10. Harry Alsop, 'Meet the Night Wolves—Putin's Hell's Angels', *The Telegraph*, 2 March 2014. http://www.telegraph.co.uk/news/worldnews/europe/russia/10670244/Meet-the-Night-Wolves-Putins-Hells-Angels.html. Accessed 7th February 2018.
11. https://www.theglobeandmail.com/news/world/tension-in-crimea-as-pro-russia-and-pro-ukraine-groups-stage-competing-rallies/article17110382/. Accessed 7 February 2018.
12. http://estonianworld.com/security/lessons-identified-crimea-estonias-national-defence-model-meet-needs/. Accessed 7 February 2018.
13. Daniel Treisman, 'Why Putin took Crimea: The Gambler in the Kremlin', *Foreign Affairs*, May–June 2016, https://www.foreignaffairs.com/articles/ukraine/2016-04-18/why-putin-took-crimea. Accessed 7 February 2018.
14. Ibid.

15. Ibid.
16. 'Crimea—The Russian Cyber Strategy to Hit Ukraine', Infosec Institute, 11 March 2014 http://resources.infosecinstitute.com/crimea-russian-cyber-strategy-hit-ukraine/#gref. Accessed 7 February 2018.
17. Ibid.
18. Can Kasapoglu, 'Russia's Renewed Military Thinking: Non-linear Warfare and Reflexive Control', Research Paper 121. http://www.ndc.nato.int/news/news.php?icode=877. Accessed 7 February 2018.

# 5

# The Mutating Wars
## 'The Hybrid Threat' in Iraq and Syria

*Shruti Pandalai*

## Introduction

For over a decade now, the United States (US) military and strategic experts, and their partners in North Atlantic Treaty Organization (NATO), have discussed threadbare the concept of 'hybrid warfare' and its relevance in understanding the changing character of war.[1] In fact, back in 2005, the then Marine Corps General and current US Secretary of Defense, James 'Jim' Mattis, co-authored an article with Frank Hoffman reflecting on the US experience in Iraq and Afghanistan. He wrote:

> We do not face a range of four separate challengers as much as the combination of novel approaches—a merger of different modes and means of war. This unprecedented synthesis is what we call Hybrid Warfare...The kinds of war we will face in the future cannot be won by focusing on technology; they will be won by preparing our people for what General Charles Krulak, the former Marine commandant, used to call the Three Block War. This is a pretty simple construct. You are fighting like the dickens on one block, you're handing our humanitarian supplies in the next block, and the next one over you're trying to keep warring factions apart...We are extending the concept a bit...We're adding a fourth block—which makes it the Four Block War. The additional block deals with the psychological or information operations aspects. The Four Block War adds a new but very relevant dimension to situations like the counterinsurgency in Iraq.

Insurgencies are wars of ideas, and our ideas need to compete with those of the enemy. Our actions in the three other blocks are important to building up our credibility and establishing relationships with the population and their leadership. The information ops component is how we extend our reach and how we can influence populations to reject the misshaped ideology and hatred they are offered by the insurgents. Successful information ops help the civilian population understand and accept the better future we seek to help build with them...All those who witnessed the Marine in Iraq understand the ultimate meaning of 'no better friend, no worse enemy.' This will be an even bigger challenge in tomorrow's Hybrid Wars, but no less relevant to victory.[2]

Since then, Mattis has repeatedly stressed the need for the 'U.S. military to transform to a "hybrid" force that expands its nonconventional means without sacrificing classic war-fighting competence.'[3] The US military thinkers took cognisance of this while planning the idea of the 'Joint Operating Environment'—a conceptual battlefield—which takes into account potential threats born out of competition for resources, economics, increased urbanisation and the possibility of non-state actors obtaining more deadly weapons, in a report called the *Joint Operating Environment 2008*.[4] A follow-on document, known as the Capstone Concept,[5] developed with the US experiences in Iraq, outlined how the US joint forces capabilities were to be implemented. It summarised its operational challenge as: 'how will future US Joint Forces with constrained resources, protect US national interest against increasingly capable enemies in an uncertain, complex rapidly changing and increasingly transparent world?'[6]

However, two crises—the emergence of terror group Islamic State of Iraq and Syria (ISIS) and its consolidation of territory in 2014 and the Russian intervention in Crimea and eastern Ukraine—provoked a renewed debate among the US and its allies on responses to hybrid threats. In February 2016, the NATO Warsaw communiqué addressed these two issues specifically, in its joint declaration: 'agreed to a strategy on NATO's role in Countering Hybrid Warfare, which is being implemented in coordination with the EU.'[7] In the Munich Security Conference in 2017, Secretary Mattis reiterated the Trump administration's commitment to its allies in the NATO and European Union (EU) in dealing with hybrid war.[8]

While some American analysts have warned that the 'idea of "hybrid warfare" in the US has become the 21st century political obsession' and 'poses a fatal threat to US security only so long as it continues to be hyperbolized',[9] others have argued that 'the introduction of hybrid warfare as a concept, albeit a vague one, was useful in nudging military strategists—as well as officials and academics—to consider more flexible and effective responses'[10] to handle this 'not so new type of war'.[11]

This chapter will discuss, analyse and draw out the application of hybrid warfare tactics by the US in Iraq and Syria in response to hybrid threats posed by the ISIS and other actors and the impact on geopolitics in the region. Since this conceptual chapter in the book covers the theoretical evolution of the concept in the US thinking, it will refrain from repeating definitions and instead look at lessons drawn by the US in these campaigns against hybrid threats; how these lessons are reflected in their military doctrines and political thinking; and if there are relevant parallels that can be drawn with the many theatres of conflicts that India finds itself mired in.

## Reviewing Conditions for Hybrid Wars: Debates beyond Hoffman's Matrix

While Frank Hoffman has defined hybrid wars as 'a tailored mix of conventional weapons, irregular tactics, terrorism, and criminal behavior in the same time and battlespace to obtain [a group's] political objectives', authors like Josef Schroefl and Stuart Kaufman have gone further to define the preconditions that help identify these conflicts which are in a constant state of mutation.[12] They argue that the current global '(dis-)order has evolved in which the international balance of state power has been largely replaced by a global equilibrium between a weary hegemon and its allies on one side, and often-violent sub-national and non-governmental actors on the other.'[13] Agreeing with Hoffman, they reiterate that 'adversaries simultaneously employ a combination of different types of warfare, including conventional, insurgent, terrorist, and cyber means.' These conflicts are characterised by their ability to 'transcend national boundaries, social and economic classes, and political ideologies' and 'blurring of lines defining enemies and allies.'[14] The adversary often has the upperhand in 'the means to surprise and spread fear throughout the traditional nation-state community, but not to establish a viable alternative order.'[15] Hence, these conflicts are protracted since 'weak states are hybridized states, governed by corrupt, predatory, and sometimes criminal elites or warlords who exercise power through patron–client ties instead of through the institutions of government.' These wars spring what they define as 'shadow globalization'[16] or 'war economies', funded by drug trafficking, extortion and kidnapping, all based on threat and violence. Yet the core of their argument remains that the approach to hybrid warfare, from the international community led by the US, has failed because of its inability 'to grapple with the core of the problem, which is not military at all, but political: the corrupt and predatory nature of the states being defended by the counterinsurgent.'[17] They make the case that:

> Hybrid wars drag on because the hybridized state is too weak to win, but usually retains enough foreign support to avoid defeat; while opposing warlords benefiting from shadow globalization profit economically as well as politically by continuing to fight. So far, Western national security policies have found no answer to this development.[18]

This argument holds true to the core case study of this chapter, which focuses on the US response to hybrid threats in Iraq and Syria, where ISIS emerged as a formidable adversary. As this book goes to print, the ISIS has been military defeated in Iraq by the US-led anti-ISIS coalition; however, as many observers have argued, 'ISIS may be on its knees but it will rise again if we don't break the cycle'.[19]

## The ISIS: The Prototype of a Hybrid Adversary

It is ironic that a 2006 report on Iraqi perspectives (published by the US Joint Forces Command), based on documents captured by coalition forces in April–May 2003, revealed that in 2003, 'Saddam Hussein had rejected the advise of a prominent general on a strategy of Iraqi warfare' that would have 'been the paradigm case of modern hybrid war, preceding Hezbollah's warfare against Israel in 2006 by three years.'[20] It described that General Raad Hamdani had proposed to Saddam Hussein an operational plan that involved 'hiding conventional Iraqi forces in cities who in combination with paramilitary Fidayeen forces would then execute, military operations from within urban centres amidst millions of civilians, from hospitals, residential areas, business streets and markets.'[21] This would involve 'launching of rockets, anti-tank and ground to air missiles, sniper attacks, hit and run tactics, terrorist bombs etc.' He also outlined the 'need to be feeding the media at home and international media with streams of selected or distorted information.'[22] This strategy had the requisite mix of military and non-military means, irregular warfare and information operations, employed in theatres simultaneously—a playbook , we now consider as the matrix for hybrid wars. Analysts argue that 'luckily for the Americans and the coalition forces, Saddam Hussein refused the advice. He dismissed the US military capabilities and overrated the strength and effectiveness of the Iraqi forces in conventional battle—to an absurd degree.'[23] If General

Hamdani had convinced Saddam, history perhaps would have been written differently.

Yet, the prototype of what the US military thinkers and their western partners describe in exhaustive detail as a hybrid adversary was the campaign waged in 2014 by the terrorist group ISIS/Da'esh in Iraq and Syria. It transitioned from being a small Iraqi affiliate of the Al Qaeda in the Arabian Peninsula (AQAP), which specialised in suicide bombings and inciting Iraq's Sunni Muslim minority against the country's Shiite majority a decade ago, to becoming a hybrid organisation which was defined as 'part terrorist network, part guerrilla army, part protostate entity'[24] that forced the US military and its allies to reassess their counter-insurgency doctrines. The ISIS drew its strength from the complex circumstances that were independently causing Iraq and Syria to fail, including domestic civil and sectarian cleavages, authoritarian leadership and polarising regional stressors.[25]

In 2014, taking advantage of withdrawal of coalition forces from Iraq, a weak government and increased sectarian violence, the ISIS marched into Iraq and captured Mosul. It used networks of ideological sympathisers, black market commerce and defensible terrain that would form the nucleus of the ISIS.[26] The ISIS quickly amassed a potent conventional army through alignment with dissident Baathist military leaders and seizure of military equipment. It defeated the Iraqi Army in a 'blitzkrieg' that unravelled four Iraqi Army divisions.[27] In June of 2014, the ISIS launched a powerful offensive on Iraq and attacked Mosul with a main striking force of 500–800 fighters deployed on Syrian soil.[28] The hybrid playbook of a mix of conventional and irregular warfare was executed by deliberately isolating, in some cases, part of the Iraqi security forces and moving towards Baghdad simultaneously from north and west. The actions were supported by a robust conventional firepower and very high mobility, leading to ISIS shortly taking control over important urban centres, lines of communication and large territories in Iraq.[29]

So, militarily, the ISIS was engaging in a campaign to amass forces to capture and hold territory. Simultaneously, it pursued political goals by declaring a caliphate, imposing its version of Sharia law on the territories it captured and globally branding its brutal jihad through its propaganda campaigns which exploited social media platforms, thus amplifying its extremist ideology by showcasing the kidnapping and execution of many local soldiers, citizens, western journalists and aid workers.[30] It employed psychological warfare to cause shock and fear among the forces facing it, and also in western countries, thus shaping the 'human terrain' for its operations. Finally, by attracting western recruits, it posed a threat in those recruits' home countries, thus making itself a transnational challenge.[31] The ISIS, hence, framed its strategy across three geographic rings: the 'interior ring' in Iraq and Syria; the 'near abroad ring' in the wider Middle East and North Africa; and the 'far abroad ring' in Europe, Asia and the US. The assessment that ISIS's strategic framework 'corresponds to a campaign with three overarching goals: to defend inside Iraq and Syria; to expand operations regionally; and to disrupt and recruit on a global scale' has been largely accurate.[32]

Frank Hoffman, in his seminal study, identified a specific set of variables that characterised hybrid threat.[33] He included: 'blended modalities' (combination of conventional and non-conventional tactics combined with terrorism and criminal activities); 'simultaneity' (employing different modes of conflict simultaneously in a coherent way); 'fusion' (threats are comprised of a mix of professional soldiers, terrorists, guerrilla fighters and criminal thugs); and 'criminality' (use of criminal activity to sustain operations and, in some cases, as a deliberate mode of conflict).

While trying to understand how the ISIS had evolved into a mutating hybrid adversary, authors Scott Jasper and Scott Moreland improvised this grid further and argued with exhaustive data on how ISIS was an adaptive, hybrid threat in transition.[34] They refined the grid as shown in Table 5.1.

**Table 5.1: Hybrid Wars: Variables, Impact, Response**

| Variables Defining Hybrid Threat | Impact on Conflict and Response |
|---|---|
| Flexible structures | Hybrid threats organise in conventional formations or distributed cells. Governance components assert control and sustain operations. |
| Terrorism | Hybrid threats utilise terror campaigns to proliferate hate and strike fear. They target cultural icons, identities and beliefs that oppose their ideologies. |
| Disregard for international law | Hybrid threats cynically view international laws as a constraint upon their adversaries that can be exploited. |
| Information warfare | Hybrid threats exploit global information access and tools to spread jihadist schemes, raise funds, recruit, train and operate. |
| Organised criminal activity | Use crime and fundraising to generate revenue to fight, govern and sustain operations. |

*Source*: Jasper and Moreland, 'ISIS: An Adaptive Hybrid Threat in Transition', n. 34.

By applying these variables, with certain improvisations, we can better understand the ISIS activity in Iraq and Syria, along with insights[35] on how hybrid threats are forcing changes in conventional military thinking and responses.

*Flexible Structures*

As the ISIS fighters captured and seized territory in Iraq and Syria in 2014, they built detailed and efficient management structures to oversee functions of finance, arms, governance, operations and recruitment. Much of their leadership rank and file included military officers from Saddam Hussein's disbanded Baathist Army, giving them an edge since they functioned like a professional army. Concurrently, al-Baghdadi developed a very efficient command and control system, which worked on the principle of 'centralised command, decentralised execution'. This translated to the ISIS also concentrating on holding ground and asserting control, apart from capturing territory. When the international coalition-led air strikes pushed back the ISIS offensive, 'the jihadists dispersed their strengths and their combat equipment in populated centres, where they mainly operated during night and re-distributed forces into small tactical elements by simultaneously limiting radio and unsecured mobile phones communications'. It was also noted that, during the same time, ISIS laid down mines and improvised explosive devices (IEDs) in order to limit, and even prevent, Iraqi and Peshmerga Kurdish forces' freedom of movement during their operations. It effectively improved its scale of operations by using vehicle-borne IEDs (VBIEDs) in order to create more chaos and breaches within the security perimeters, hence providing the opportunity for suicide attackers or armed fighters to inflict more casualties. Among its arsenal, the use of drones for air reconnaissance missions showcased its adaptability to new technologies. It was also reported that ISIS used chemical agents during the offensive launched against the Iraqi city of Dhuluiya in October 2014. The ISIS was able to employ simultaneously many elements of kinetic and non-kinetic warfare, such as conventional army tactics, urban guerrilla warfare, network-centric warfare, subversion and chemical, biological, radiological and nuclear (CRBN) warfare, to boost its operational success against its adversary.

The ISIS had gained major advantages mostly due to high manoeuvrability and ability to surprise the adversary during conventional warfare operations. These were achieved through marches and quick deployment of forces, using high-quality and dense land communication networks, culminating with night or early morning raids. At the same time, reconnaissance, small engagements and diversion tactics were repeatedly used in the ISIS operations in order to identify security forces readiness and reaction procedures, avoid their strong points and divert them from the main target. In what can be identified as the use of a traditional kinetic tool to achieve a non-kinetic objective, occasionally armoured vehicles were used as part of the psychological warfare conducted by the ISIS, which was aimed at surprising the enemy and inflicting on it a profound feeling of fear and helplessness. After the ISIS took control of Iraq's Jalula city in August 2014, it organised a defensive establishment comprising of T-55 and T-62 tanks, anti-tank weapons and recoilless guns, in order to protect ground avenues of approach towards the city. It also reportedly managed to obtain air defence capabilities by taking over weapon depots and was said to have used air defence systems such as ZU-23s, FN-6, SA-7s and SA-16s while defending its captured territories.

The ISIS's ability to attract and recruit foreign fighters, as well as absorb local militia comprised of Iraqi Sunni insurgent groups, provided it formed military units that could operate and maintain captured Iraqi and Syrian Army equipment and fight as a disciplined and organised force. These militia acted not just as a defensive force but had the advantage of being backed by a sympathetic civilian populace. Here, again, the ISIS was able to exploit the combination of insurgents and proxies and gained from a sustained campaign of ideological warfare to draw support from the local population.

*Terrorism*
Cruel acts of terrorism were often used by the ISIS to subdue local populations and ensure compliance. From razing Shiite shrines to executing defectors, overrunning security forces and hoisting its black flag above government buildings, its highly publicised brutality on social media included public beheadings and targeting of cultural icons and religious centres with an aim to erase history. This deliberate targeting of cultural icons has been referred to as the 'weaponisation of culture' and was employed to maximum effect by the ISIS.[36] Simultaneously, the ISIS was also accused of committing genocide in an effort to exterminate the Yazidi religious minority in Syria and Iraq, according to the United Nations (UN) investigators. After losing ground militarily, the ISIS shifted its course to extra-regional targets. Various attacks across Europe, the US and South and Southeast Asia, seen as ISIS-inspired lone wolf attacks, point to the ability, adaptability and mutation of the ISIS to become a transnational threat.

## Disregard for International Law

The ISIS, through its much publicized mass executions of minority groups and captured soldiers, journalists, etc., demonstrated disregard for internationally accepted laws and universal human rights. A UN panel report, widely cited by the press, affirmed that ISIS fighters raped, sexually mutilated and sterilised Yazidi women to prevent the birth of their children. The reported rape of non-Muslim women and children as young as 9–12 years in ISIS camps and captured territories was defended as a religious duty in its publications.[37] Witnesses attested to the fact that this philosophy of annihilation extended even to hospitals in liberated towns that were booby-trapped with explosives. The ISIS even defended the use of children, the so-called 'cubs of the caliphate', to fill gaps in fighters and suicide bombers.

## Information Warfare[38]

The rise of ISIS as the poster child of global jihad and its unprecedented visibility was largely due the strategic storytelling it employed with the use of social media. The scale and quality of the ISIS's content was unmatched. A Brookings report studying counter-narratives to ISIS propaganda stated:

> Efforts to blunt ISIS propaganda have been tentative and ineffective, despite major efforts by countries like Saudi Arabia, the United States and the United Kingdom, and even al-Qaida. Counter-messaging efforts, such as those by the U.S. Department of State's Center for Strategic Counterterrorism Communications (CSCC), have been dwarfed by the sheer size of the ISIS footprint.[39]

The report further added, 'no effort to date has matched the tailored nature, the scope, nor the electrifying content of the Islamic State's material.'[40] Even after its military defeat, the group reportedly was able to churn out approximately 20 media products each day.[41] The emphasis on media messages to target both followers and adversaries was a prerequisite to ensure that the idea of ISIS withstands even as its territorial gains declined. It was also noted that during a single summer month, ISIS produced nearly 900 pieces of Arab-language propaganda, and nearly half focused on quality of life issues, such as food, utilities and schools, in an attempt to portray a utopian view of life under the caliphate.[42] In the same year, the ISIS decided to not just exploit the Internet for propaganda purposes but use it as a weapon. The ISIS sympathisers hacked the Twitter and YouTube feeds at the US Central Command to publish lists of Generals and addresses along with videos, including the title 'Flames of War'.[43]

## Resource Warfare and Organised Criminal Activity[44]

It is no secret that during the height of its success in 2014, the ISIS raked in over $2 billion from seized oilfields in its captured territories, looted banks and used extortion of people under its control as means to fill its coffers. The ISIS, within its 'protectorate', taxed everything from income to bank transactions to education of school children to generate over $300 million a year. It exerted control over essential resources, including staples like wheat and barley, core industries like mines and cement factories, and of course oil, which accounted for over $500 million a year in 2014. It is said that the ISIS 'regional operations thrived because they were able to make conquest profitable.'[45] However, in 2015, the fall of Ramadi, Fallujah and Kirkuk meant the loss of lucrative oilfields and this forced the ISIS to increase taxes in their protectorate. This was followed by looting of cities that it was forced to abandon and plundering of priceless artefacts from museums and archaeological sites. The ISIS is said to have developed an extensive antiquities smuggling enterprise, which it covered up by deflecting the world's attention through wanton destruction of historical treasures. It has been noted by analysts that while on the face of it, the ISIS made a big spectacle of statues, tombs and temples being destroyed with sledgehammers and bulldozers, behind the scenes they smuggled smaller items such as figurines and masks to European buyers via black market transit routes. It has been estimated by Iraq's State Board for Antiquities and Heritage that the income from the activities described was millions of dollars.

## The US Response to ISIS as a Hybrid Adversary

The Obama administration led a multilateral coalition in support of its stated goal to 'degrade and ultimately destroy'[46] the ISIS organisation, with an aim to progressively reduce the geographic and political space, manpower and financial resources available to it. The strategy, now famous as the Obama Doctrine 2014, outlined the ISIS as a hybrid threat and focused on a number of 'lines of effort', including, 'in partnership with several European and Arab states: direct military action, support for Iraqi and Syrian partner ground forces, intelligence gathering and sharing; and efforts to restrict flows of foreign fighters, disrupt the Islamic State's finances and eliminate its leaders.'[47]

In Iraq, the Obama administration planned to: support security forces under central government command; maintain support for forces affiliated with the Kurdistan Regional Government; and preserve Iraq's political and territorial unity pursuant to its constitution.[48] The international coalition against ISIS responded to an adaptive hybrid adversary via a combination of traditional air power, weapons supplies to Kurdish Peshmergas, the deployment of advisors to Iraqi government troops and sectarian militias and training activities for opposition forces.[49]

In terms of military operations, the US and coalition forces used combat aircraft, armed unmanned aerial vehicles (UAVs) and sea-launched cruise missiles to conduct more than 17,000 strikes against ISIS targets in Iraq and Syria in the initial six weeks of the campaign from 8 August 2014.[50] According to the US Congress research report, till October 2016, the US had already spent $10.06 billion in kinetic operations.[51] The report said that currently, the US 'strikes support defensive and offensive military operations by Iraqi military and Kurdish forces in Iraq and seek to weaken the ISIS organization's control over its remaining strongholds inside Syria.'[52]

Yet, typical to characteristics identified as variables of hybrid states, the prospects and options for undermining the ISIS supporters have been shaped by the relative success or failure of efforts to restore security, boost economic growth, address political grievances and promote effective governance. The reconstruction and rehabilitation efforts will require double the effort in lieu of the power vacuum left behind in the post-ISIS period. According to their own internal assessments, over the longer term, two durable challenges confront the US and international community from the campaigns launched by the ISIS.[53] The first challenge is the result of the blurring of lines between allies and adversaries due to the mobilisation of armed groups for combating the ISIS. This complicates efforts to resolve political disputes over the governance of areas recently freed or captured from the ISIS in various countries. It is particularly problematic since future conflict cannot be ruled out among groups who were superficially joined in opposition to the ISIS. Second is the social challenge posed by the mobilisation of the thousands of individuals who have travelled to various battlefields in support of the ISIS and other extremist groups in recent years. For, as the US Secretary of State Rex Tillerson said, 'defeating the Islamic State globally may be "extremely challenging"' and 'depriving the group of its so-called caliphate in the Middle East "will not defeat ISIS once and for all, it will simply morph to its next version."'[54]

The chain of events that fuelled the rise of the ISIS and the conditions which helped its expansion show no sign of abating. Complicating matters are the scope and nature of Russian, Iranian, Turkish, European and Arab involvement in Syria and Iraq, which have major impact on the context in which the US leaders will formulate strategy. The next section takes on the impact of evolving geopolitics in dealing with hybrid threats and challenges for the region.

## Syrian Civil War as a Hybrid War

So far, this chapter has analysed how the ISIS became the prototype of the hybrid threat for the US in Iraq. However, it is the civil war in Syria, which broke out in 2011, which provided the crucible for the resurgence of the ISIS in Iraq and Syria in 2014. The Syrian Civil War is now being seen as a protracted hybrid war, where the ISIS is just one actor on the larger chessboard. Syria is beginning to resemble the weakened hybridised state, where hundreds of different groups—with different objectives and different connections to the outside—fight and compete for the same turfs and the same spoils. Some of these groups managed to merge into what is now the ISIS. Although the conflict in Syria started as an internal uprising, the 'revolution' currently appears to have been co-opted by international players, namely, the US, Russia, Iran, Turkey and the Arab states. While both the US and Assad regime have targeted the ISIS and its elimination from Syria as their mission objectives, the reality is more nuanced. Observers argue that while the ISIS is usually said to be well organised, well armed, and well funded, the backlash launched by previously allied militant groups against the ISIS in the beginning of January 2014 is evidence that its prominence should not be conflated with leadership. While the revised hybrid threat grid remains a valid mode for assessment of the ISIS ability to expand and hold territory in Syria, in this case it is a much smaller part of the larger game being played in the troubled region. The next section looks at the various actors and interests at play in Syria and how they confine themselves to the characteristics of a hybrid war.

## Actors and Interests in Syria's Hybrid War[55]

### *Discontent with Bashar al-Assad Regime*

In 2011, 15 school children were arrested and tortured for writing anti-government graffiti. The use of force by the Assad regime, against non-violent protests, triggered unrest and rebellion in various provinces. The opposition movement eventually veered into two umbrella groups—one political, one armed—both based primarily in exile. Political groups merged to form the Syrian National Council (SNC), although members struggled to establish trust and develop shared goals. A small number of junior military defectors formed the Free Syrian Army (FSA), which claimed leadership over the armed opposition but whose authority was generally unrecognised by local armed groups.

After six years of conflict, the challenges posed by the situation in Syria have multiplied and evolved. The country's descent into brutal war has created a multifaceted regional crisis, marked by the mass displacement of civilians, the emergence and empowerment of violent armed Islamist extremist groups, gross human rights abuses and war crimes, the use of chemical weapons, the proliferation of arms and the covert and overt intervention of outside actors. In short, a textbook case for variables that characterise protracted hybrid warfare (as explained in Table 5.2).

Since late 2015, Assad and his government have leveraged military, financial and diplomatic support from Russia and Iran to improve and consolidate their position relative to the range of anti-government insurgents arrayed against them. These insurgents include members of the ISIS, Islamist and secular fighters and Al Qaeda-linked networks. While the ISIS forces have lost territory to the Syrian government, to Turkey-backed Syrian opposition groups and to the US-backed Syrian Kurdish and Arab fighters since early 2016, they remain capable and dangerous. The ISIS 'capital' at Raqqah has been isolated and liberated, but large areas of eastern Syria remain under the group's control. As this book goes to print, news has emerged of the last of the ISIS fighters in Raqqah brokering a deal to leave the Syrian city 'as they run out of ammunition after three-month battle with the US-led coalition backing the SDF's drive'.[56] The presence and activities of Russian military forces and Iranian personnel in Syria create complications for the US officials and military planners, and raise the prospect of inadvertent confrontation with possible regional or global implications. The following sections elaborate on differing interests, actors and their role in hybridising the war in Syria.

Table 5.2: Snapshot of Actors and Conflicting Interests in Syria's Hybrid Wars[57]

| | |
|---|---|
| Turkey | Staunch supporter of the rebels. It backs US-backed Syrian Democratic Forces (SDF) alliance against the ISIS, but is against Kurdish groups being armed by the US. It allowed a southern border with Syria so porous that it offered the ISIS and the Al Qaeda-affiliated Jabhat al-Nusra a logistical safe zone. |
| Iran | The Shia power has spent billions to bolster the Alawite-dominated government, providing military advisers and subsidised weapons, as well as lines of credit and oil transfers. It is also widely reported to have deployed hundreds of combat troops in Syria. Syria is the main transit point for Iranian weapons shipments to the Lebanese Shia Islamist movement, Hezbollah, which has sent thousands of fighters to support Assad's forces. |
| Saudi Arabia and Qatar | Jockeying for regional influence and wanting to hit back against rival Iran. They are the major providers of military and financial assistance to the rebels, including those with Islamist ideologies, providing easy recruiting grounds for the extremists. |

| | |
|---|---|
| Russia | Assad's ally has blocked UN resolutions against the regime and supplies weapons to the Syrian military. Moscow wants to protect a key naval facility which it leases at the Syrian port of Tartous, which serves as Russia's sole Mediterranean base for its Black Sea Fleet, and has forces at an air base in Latakia, President Assad's Shia Alawite heartland. In September 2015, Russia launched air strikes against rebels but Western-backed groups were reported to have been hit. |
| ISIS | It had captured swathes of territory in northern and eastern Syria, but is battling government forces, rebel brigades and Kurdish militias, as well as facing air strikes by Russia and a US-led multinational coalition. |

*Power Brokers: The Role of Russia, Iran and Turkey in Syria's Hybrid War*
On 22 November 2017, the heads of state of Russia, Iran and Turkey attended a trilateral summit in Sochi to coordinate their efforts on the future of Syria. Moscow had declared its plans to host a national Syrian congress in Sochi, consisting of different Syrian factions, to discuss the political process as well as a new constitution for the country.[58] The meetings have shown that the Russia–Iran–Turkey trio are really the power brokers that may decide the future of Syria.

In the case of Russia, analysts believe that:

> initiating various meetings with their partners in the Astana process, namely Iran and Turkey, and at the same time underlining the importance of Iran and Russia's military support for the Syrian government in shaping developments on the ground, the Russians want to be recognized as the main pillar of military success over the Islamic State (IS) and other terrorists. Moscow's frequent criticism of the US-led anti-IS coalition, accusing it of not being serious in fighting terrorism, could be interpreted in the same vein. In other words, Russia is trying to say that without its efforts there would be no end in sight to the rule of the United States, nor would it be possible to forge a compromise between Iran and Turkey's military plans in Syria.[59]

Observers say:

> [by] proposing its initiative for the national Syrian congress, Moscow is also seeking to have final say in the domestic equations in Syria during the transitional period ahead. While Turkey's objection to the inclusion of Syrian Kurdish groups in the political process as well as the hard-line stances of some opposition groups regarding the role of Assad and his government in the future of Syria, have

forced Moscow to delay the congress, Russia's actions show a willingness to act as a mediator between different Syrian factions.[60]

Russia realises that with the war ending and the focus shifting to reconstruction of post-war Syria, others, including China, Europe and Japan, will step forward. Moscow thus seeks to secure itself 'a piece of the lucrative reconstruction effort',[61] which will be financed by international donors. It is also looking to secure its own core interests in Syria, whatever the balance of political power in the country, says an assessment in *Foreign Affairs*:

> Among these is a permanent air and naval presence in the country. Under the lease agreements signed in 2015 and 2016 with Damascus, both the Khmeimim air force base and the Tartus naval facility, which is being upgraded to a regular naval base, will stay in place for decades after the end of the war. The Syrian armed forces will continue to rely on Russian weapons and equipment, and Russian military specialists will continue to advise and train their Syrian colleagues. This will seal Syria's role as Russia's main geopolitical and military foothold in the Middle East.[62]

Since September 2015, Iran and Russia's military support—boots on the ground and air power, respectively—has ensured the al-Assad regime's survival in Syria.[63] However, this alliance is not without differences, which led Russia to support Turkey's military incursion into northern Syria in order to provide Moscow with a stronger hand in shaping a political settlement to the war.[64] Turkey and Iran's competing regional interests allow Russia to serve as the broker between them because Russia enjoys better relations with both Turkey and Iran than they enjoy with each other. Russia is concretising its strategic military gains with the Assad regime, while limiting both Iran and Turkey's ability to shape outcomes in Syria without Russia's consent.[65]

Meanwhile, Tehran wants to institutionalise its presence on the ground in Syria after the war, both to influence the future of that country and to maintain a physical link to its main regional ally, Hezbollah.[66] Militarily, for Iran, the Syrian war has been transformational,[67] as it has proved that it has the capacity to conduct quasi-conventional warfare hundreds of miles from its borders. According to analysts, this capability, which very few states in the world have, 'will fundamentally alter the strategic calculus and balance of power within the Middle East.'[68] Experts argue that Iran's constant reworking of its hybrid model of warfare in Syria:

will strengthen its capacity to project power and will facilitate Tehran's efforts to deploy forces alongside similar proxy forces in other theaters, such as Iraq or Lebanon and to expand and improve the capabilities of its proxies and direct them against US interests and allies if it chooses.[69]

Close cooperation between Russian and Iranian military personnel at the operational and tactical levels in Syria is also 'introducing Iran and its proxies to signature Russian campaign-design concepts such as cauldron battles, multiple simultaneous and successive operations, and frontal aviation'.[70] The knowledge transfer between Iran and Russia presents, according to western pundits, 'the US with a more capable Iran that remains hostile to the US and its allies in the region' and is game changer in the 'balance of power'.[71]

In the case of Turkey,[72] the primary goal is to prevent formation of an autonomous Kurdish region on its borders with Syria and Iran. It is no secret that President Erdogan's Turkey has been supporting elements of the Syrian opposition for the past five years and has been angling to bring them into the political process. Turkey's ties to the Syrian opposition, which Russia views as an asset in its efforts to broker a political solution, are viewed with deep scepticism if not outright hostility by Iran.[73] Russia and Turkey's 'marriage of convenience' appears to serve both of their interests in Syria. According to experts:

> Turkey obtains a free hand in northern Syria to stymie the territorial ambitions of the U.S.-supported Syrian Kurds, while Russia gets a compliant partner that will constrain Iran's influence and help wind down the conflict by co-opting portions of the 'revolutionary' (as opposed to jihadist) anti-Asad Sunni opposition...Turkish military forces on the ground in northern Syria, supported by Russia air power, also provide Moscow with the means to limit the ambitions of both the Kurds and Iran in Syria, without expending a great deal of Russian blood or treasure.[74]

It also helps Russia demonstrate its new leverage as the arbitrator between Shia Iran and Sunni Turkey in Syria.

It is clear that Syria has become the ground for exercising and securing strategic influence and power for Russia, Iran and Turkey through all means necessary.[75] However, there is concern that 'Moscow's attempt to centralize the political process around its own role could potentially alienate Iran down the road, thereby challenging their so-far-successful partnership in Syria'[76] and pushing the region into further uncertainty. The Russian handling of the Syrian Civil War has left the US and its allies very little room for manoeuvre. The next two sections deal with the US dilemmas and response to the crisis in Syria.

*The US Response and Policy Initiatives in Syria*[77]

The central dilemma for US policy on Syria has been on the question of whether or not it should support and pursue a resolution of the conflict that would recognise a continuing role for Syrian President Bashar al-Assad. According to the numerous assessments put up by the US Congressional Research Service,[78] policy proposals over time have reflected "various assumptions" on the outcomes of either Assad's rule or fall would have on the ending of the conflict. Concerns over regional stability and counter-terrorism efforts have been amplified by the developments in Syria's conflict and perhaps trumped US' concerns about Assad's future. The perpetual dilemma confronted by US and regional policymakers has been of which problem to confront first—Assad or the ISIS. They have tried to also debate on the various approaches and their outcomes would affect Syria's stability, both in the short- and long-term.

A June 2013 report of the US Congressional service notes:[79]

> "President Obama and his Administration have been calling for Asad's resignation since August 2011, and have pressed the United Nations Security Council to condemn the Syrian government. The United States has recognized the National Coalition of Revolution and Opposition Forces (SC) as the legitimate representative of the Syrian people and has provided nonlethal assistance to the Coalition and an affiliated Supreme Military Council (SMC). The Obama Administration believes that a negotiated political settlement is required and has prepared military plans to secure Syria's stockpiles of chemical weapons, if necessary. (....) (However) Some observers advocate for more robust civil and military aid to the SC and SMC as a means of forcing the Assad regime to the negotiating table. Opponents of this approach argue that making opposition groups more formidable could intensify the fighting and risks empowering extremists. Some armed opposition factions, including powerful Islamist coalitions, reject negotiation."

The policy dilemmas are aplenty, since US strategy has not yielded the desired outcomes in some cases. For example, since its intervention in Syria in 2014, locally; the US has backed Syria's main opposition alliance, the National Coalition, and has provided limited military assistance to 'moderate' rebels. However according to media reports, in 2011, the US provided the Free Syrian Army with non-lethal aid (including food rations and pickup trucks), but subsequently began providing training, cash and intelligence to selected Syrian rebel commanders.[80] While the Central Intelligence Agency (CIA)-run programmes (special ops) to arm and train Syrian opposition factions began in 2013, on 17 September 2014, the House of Representatives voted to authorise the executive branch to train and equip Syrian rebels against the ISIS forces.[81] However, the programme to train and arm 5,000 Syrian rebels to take the fight to ISIS on the ground suffered deep setbacks, with few having even reached the frontline.[82]

The CRS assessment notes that "while equipment losses have not proven to be a major systemic concern since the change was announced, some Syrian opposition groups that reportedly have received U.S. equipment and weaponry have surrendered or lost these items to other groups, including to the Islamic State."[83] It makes a point that initially, the "comprehensive training approach sought to create unit cohesion, groom and support reliable leaders to serve as U.S. partners" going beyond, "local, sectarian, or ideological goals."[84] However over a period of time, the process "effectively equipped some anti-IS forces in some areas of Syria" and has "had less quantifiable effects on the development and practices of opposition forces that may influence security in Syria for years to come."[85]

Meanwhile, President Trump and his administration have identified the defeat of the ISIS organisation as their highest priority in the Middle East, and he has directed his administration in January to develop 'a new plan to defeat ISIS'.[86] A major policy concern of the US has been the use or loss of control of chemical weapons in Syria during the ongoing civil war. According General Joseph Votel, Commander of U.S. Central Command, CENTCOM, "ISIS' use of chemical weapons and its evolving application of available off-the-shelf technologies that include unmanned aerial systems now used for both observation and to achieve lethal effects, poses a growing threat. For example, ISIS has reportedly used chemicals, including sulfur mustard and toxic industrial chemicals, in attacks more than 50 times in Iraq and Syria since 2014"[87] The reported use of the nerve agent Sarin by aerial bombardment on 4 April 2017, in the town of Khan Sheikhoun in rebel-held Idlib province, killed an estimated 80–100 people and brought about a US military response including US defensive strikes on Syrian government forces.[88] This development complicated the Trump administration's attempts to pursue a more cooperative relationship with the Russian government which is a major player in this conflict. Contrarily, US efforts to contain, or limit Iran's security support for its foreign partners also had implications for the war in Syria. In essence, the US policy in

Syria would be best described as a two staged process – with the first focusing on the defeat of the ISIS and the second on the 'de-escalation' of violence in Syria, including through ceasefire agreements.[89]

*Dilemmas in the US Policy Response: Preparing for a Protracted Hybrid War*
For the US, President Assad and Russia's refusal to accept what they describe as Western-led regime change in Syria and case for a counter-terrorism cooperation with the Syrian government against its adversaries as a pre-condition for transition arrangements, remain unviable options. The fallout in terms of a threat of direct confrontation with Russia, cannot be ruled out if efforts to compel Assad's departure are realized. These decisions have broad implications beyond Syria. At the same time, the risk remains that any perceived US acknowledgement to or cooperation with Russia's intervention on Assad's behalf risks alienating anti-Assad forces and their regional backers, as well as providing Russia with an opportunity to consolidate a new, active role for itself in regional security arrangements.

The US Congressional report assessment notes:

> Over the longer term, Syria's diversity and the interplay of its conflict and regional sectarian rivalries raise the prospect of continued violence even in the wake of the type of 'managed transition' that has at times been identified as a U.S. policy goal. The presence and power in Syria of armed groups directly opposed to the governance models promoted by many Syrians and the United States suggests that the conflict could persist after any negotiated settlement seeking to replace the current Assad-led government with a government of national unity or other inclusive formulation. Political opposition coalitions active internationally appear to lack grassroots support and, because of their lack of material control over the most powerful armed groups. They appear to lack the ability to guarantee security commitments that might presumably be part of a negotiated settlement. Some analysts doubt the Assad government could survive a partial transition and suggest state collapse could accompany efforts to replace it whether by negotiation or by force. Even under relatively favorable circumstances, state weakness may allow extremist and terrorist groups to operate from Syria for years to come.[90]

This is a bleak outlook and reinforces the case for looking at the Syrian Civil War as a protracted hybrid war that US military planners need to constantly adapt themselves to.

## Lessons from the US Experience

- The debate in the US on hybrid wars has prompted military and academic thinking to go beyond the binaries of war and peace.
- Hybrid threat responses have a playbook, where actions range across strategic and tactical dimensions; from seizing of financial assets to limiting the movement of extremists. Regionally led military counter-offensives, closure of borders, disruption of financing, prosecution of atrocities, protection of persecuted minorities and prevention of mass media exploitation for recruiting and training, all are parts of the strategy.
- The ISIS maybe losing militarily but as an adapting hybridised force, it is likely draw upon low-profile tactics now that it is faced with a strong anti-ISIS coalition in Iraq. Hybridised warfare gives the ISIS resilience and flexibility to adapt and evade defeat. This is a lesson for other insurgencies that are evolving across the world.
- The destruction of the ISIS's physical caliphate can contain the threat only for a period of time, unless effective states in Iraq and Syria arise to prevent its return. Reconstituting these states is necessary to diminish the sectarian polarity of the Middle East, already charged by the proxy war between Arab states and Iran that is evident in Syria and Yemen.
- Regional sectarianism inhibits responses to hybrid threats: It has been observed that the social mobilisation of the Arab world against Assad and Iran gave both the ISIS and Al Qaeda greater freedom of action. Iran is a higher priority than ISIS for many Arab states that are members of the anti-ISIS coalition. Analysts believe, these states are prioritising military action to contain and push back Iran and its proxies in Syria and Yemen over anti-ISIS action.[91]
- Finally, the concepts of information warfare are severely lacking responses in military thinking even in countries like the US. It would be important to fight back in the battle of the narrative, countering powerful propaganda messages. But this is challenging for democratic governments, posing particular issues regarding the use of modern media. The NATO and its member governments are, therefore, putting new emphasis on strategic communications. It is a lesson India would do well to take to also.

As Lieutenant General Michael Vane, US Army, wrote decisively in 2011:

Specialized skills are essential for successful operations. The specialized skills required of soldiers today and in the future are articulated as 'New Norms'. They include operational adaptability, cultural and language proficiency, negotiation, digital literacy and space knowledge, weapons technical intelligence, and site exploitation. These specialized skills must now become universal tasks.

## NOTES

1. For more, see Frank Hoffman, 'Conflict in the 21st Century: The Rise of Hybrid Wars', Potomac Institute for Policy Studies, December 2007, available at http://www.potomacinstitute.org/publications/23-publications/reports/1267-conflict-in-the-21st-century-the-rise-of-hybrid-wars, accessed on 20 August 2017; Christopher Paul, 'Confessions of a Hybrid Warfare Skeptic', *Small Wars Journal*, 3 March 2016, available at http://smallwarsjournal.com/jrnl/art/confessions-of-a-hybrid-warfare-skeptic, accessed on 25 August 2017; Josef Schroefl and Stuart J. Kaufman, 'Hybrid Actors, Tactical Variety: Rethinking Asymmetric and Hybrid War', *Studies in Conflict & Terrorism*, Vol. 37, No. 10, 2014, pp. 862–880; and Damien van Puyvelde, 'Hybrid War—Does it even Exist?', *NATO Review*, 2015, available at http://www.nato.int/docu/review/2015/also-in-2015/hybrid-modern-future-warfare-russia-ukraine/EN/index.htm, accessed on 25 August 2017.
2. James N. Mattis and Frank Hoffman, 'Future Warfare: The Rise of Hybrid Wars', *Proceedings*, Vol. 132, No. 11, November 2005, available at https://www.usni.org/magazines/proceedings/2005-11/future-warfare-rise-hybrid-wars, accessed on 1 September 2017.
3. 'General Mattis: U.S. Must Prepare for "Hybrid" Warfare', *Small Wars Journal*, 13 February 2009, available at http://smallwarsjournal.com/blog/general-mattis-us-must-prepare-for-hybrid-warfare, accessed on 1 September 2017.
4. Ibid.
5. 'Capstone Concept for Joint Operations (CCJO): Joint Force 2020', September 2012, available at http://www.dtic.mil/doctrine/concepts/ccjo_jointforce2020.pdf, accessed on 1 September 2017.
6. Ibid.
7. 'Warsaw Summit Communiqué: Issued by the Heads of State and Government Participating in the Meeting of the North Atlantic Council in Warsaw 8–9 July 2016', available at http://www.nato.int/cps/en/natohq/official_texts_133169.htm, accessed on 1 September 2017.
8. Remarks by the US Secretary of Defense, James Mattis, at the Munich Security Conference in Munich, Germany, 17 February 2017, available at https://www.defense.gov/News/Speeches/Speech-View/Article/1087838/remarks-by-secretary-mattis-at-the-munich-security-conference-in-munich-germany/, accessed on 1 September 2017.
9. Rebecca Farrar, 'Understanding the Role of Hybrid Warfare and U.S. Strategy for Future Conflicts', Centre for International Maritime Security, 23 April 2017, available at http://cimsec.org/understanding-role-hybrid-warfare-u-s-strategy-future-conflicts/32171, accessed on 1 September 2017.

10. 'Countering Hybrid Threats: Challenges for the West', *Strategic Comments*, Vol. 20, No. 8, 2014, pp. x–xii.
11. Frank Hoffman, 'On Not-so-New Warfare: Political Warfare vs Hybrid Threats', *War on the Rocks*, 28 July 2014, available at https://warontherocks.com/2014/07/on-not-so-new-warfare-political-warfare-vs-hybrid-threats/, accessed on 1 September 2017.
12. For more, see Hoffman, 'Conflict in the 21st Century', n. 1, p. 29 (pp. 29–35); and Schroefl and Kaufman, 'Hybrid Actors, Tactical Variety: Rethinking Asymmetric and Hybrid War', n. 1, doi: 10.1080/1057610X.2014.941435, accessed on 1 September 2017.
13. Ibid.
14. Ibid.
15. Ibid.
16. Ibid.
17. Ibid.
18. Ibid.
19. Hassan Hassan, 'Isis may be on its Knees but it will Rise Again if We don't Break the Cycle', *The Guardian*, 15 July 2017, available at https://www.theguardian.com/world/2017/jul/15/isis-rise-again-defeat-mosul-islamic-state, accessed on 1 September 2017.
20. Kevin M. Woods et al., *The Iraqi Perspectives Report : Saddam's Senior Leadership on Operation Iraqi Freedom from the Official US Joint Forces Command Report*, Annapolis, MD, 2006, quoted in Øyvind Østerud, 'The Challenges of Hybrid Wars', The Norwegian Atlantic Committee Security Brief 2-2015, Department of Political Science, University of Oslo, available at http://www.atlanterhavskomiteen.no/files/dnak/Documents/Publikasjoner/2015/KortInfo/SB%202-15.pdf, accessed on 2 September 2017.
21. Ibid.
22. Ibid.
23. Ibid.
24. Ralph D. Thiele, 'The New Colour of War—Hybrid Warfare and Partnerships', ISPSW Strategy Series: Focus on Defense and International Security, Issue No. 383, October 2015, available at https://www.files.ethz.ch/isn/194330/383_Thiele.pdf, accessed on 2 September 2017.
25. Ibid.
26. Ibid.
27. Ahmed Salah Hashim, 'State and Non-state Hybrid Warfare', *Sustainable Security*, 30 March 2017, available at https://sustainablesecurity.org/2017/03/30/state-and-non-state-hybrid-warfare/, accessed on 2 September 2017.
28. Liviu Ionita and Beng Iulian Aanitei, 'Elements of Hybrid Warfare in the Islamic State of Iraq and Levant Operations', *Romanian Military Thinking*, Vol. 2, April–June 2015, pp. 28–39, available at http://smg.mapn.ro/gmr/Engleza/Ultimul_nr/ionita,aanitei-p.28-39.pdf, accessed on 2 September 2017.
29. Ibid.
30. 'Countering Hybrid Threats', n. 10.
31. Ibid.
32. Jessica Lewis McFate, 'The ISIS Defense in Iraq and Syria', Middle East Security Report No. 27, May 2015, available at http://www.understandingwar.org/sites/

default/files/ISIS%20Defense%20in%20Iraq%20and%20Syria%20—%20Standard.pdf, accessed on 1 September 2017.
33. Frank G. Hoffman, 'Hybrid vs. Compound War: The Janus Choice—Defining Today's Multifaceted Conflict', *Armed Forces Journal*, 1 October 2009, available at http://www.armedforcesjournal.com/hybrid-vs-compound-war/, accessed on 1 September 2017.
34. Scott Jasper and Scott Moreland, 'ISIS: An Adaptive Hybrid Threat in Transition', *Small Wars Journal*, 29 October 2016, available at http://smallwarsjournal.com/jrnl/art/isis-an-adaptive-hybrid-threat-in-transition#_edn5, accessed on 2 September 2017.
35. The following subsections have inferences drawn primarily from two research articles: Jasper and Moreland, 'ISIS: An Adaptive Hybrid Threat in Transition', n. 34; and Ionita and Aanitei, 'Elements of Hybrid Warfare in the Islamic State of Iraq and Levant Operations', n. 28.
36. For more, see Rochelle Davis, 'Culture as a Weapon System', *Middle East Report*, No. 255, 2010, pp. 8–13, available at www.jstor.org/stable/40660865, accessed on 20 September 2017.
37. Rukmini Callimachi, 'ISIS Enshrines a Theology of Rape,' *The New York Times*, 13 August 2015. accessed at https://www.nytimes.com/2015/08/14/world/middleeast/isis-enshrines-a-theology-of-rape.html on 2 Sep 2017
38. Information for this section has been sourced from Colin Clarke and Charlie Winter, 'The Islamic State may be Failing, but its Strategic Communications Legacy is Here to Stay', *War on the Rocks*, 17 August 2017, available at https://warontherocks.com/2017/08/the-islamic-state-may-be-failing-but-its-strategic-communications-legacy-is-here-to-stay/, accessed on 2 September 2017. Also see Jasper and Moreland, 'ISIS: An Adaptive Hybrid Threat in Transition', n. 34.
39. Alberto M. Fernande, 'Here to Stay and Growing: Combating ISIS Propaganda Networks', The Brookings Project on US Relations with the Islamic World, US–Islamic World Forum Papers, 2015, p. 1, available at https://www.brookings.edu/wp-content/uploads/.../IS-Propaganda_Web_English.pdf, accessed on 2 September 2017.
40. Ibid.
41. Jasper and Moreland, 'ISIS: An Adaptive Hybrid Threat in Transition', n. 34.
42. Ibid.
43. Ibid.
44. Ibid.
45. Ibid.
46. 'Statement by the President on ISIL', Office of the Press Secretary, The White House, 10 September 2014, available at https://obamawhitehouse.archives.gov/the-press-office/2014/09/10/statement-president-isil-1, accessed on 2 September 2017.
47. Christopher M. Blanchard, 'The Islamic State and U.S. Policy', Congressional Research Service Report, February 2017, available at https://fas.org/sgp/crs/mideast/R43612.pdf, accessed on 2 September 2017.
48. Ibid.
49. Ibid.
50. Ibid.
51. Ibid.

52. Ibid.
53. Ibid.
54. Ibid.
55. Information for this section has been sourced from Carla E. Humud, Christopher M. Blanchard and Mary Beth D. Nikitin, 'Armed Conflict in Syria: Overview and U.S. Response', Congressional Research Service Report, 10 August 2017, available at https://fas.org/sgp/crs/mideast/RL33487.pdf, accessed on 2 September 2017.
56. Damien Gayle, 'Last ISIS Fighters in Raqqa Broker Deal to Leave Syrian City—Local Official', *The Guardian*, 14 October 2017, available at https://www.theguardian.com/world/2017/oct/14/last-isis-fighters-in-raqqa-seek-deal-to-leave-former-capital-in-syria, accessed on 2 November 2017.
57. Information in this section is culled out from : Zachary Laub, Backgrounder , "Who's Who in Syria's Civil War", Council on Foreign Relations, 28 April 2017, accessed on URL: https://www.cfr.org/backgrounder/whos-who-syrias-civil-war, on 2 September 2017 and "Syria crisis: Where key countries stand" www.bbc.com , 30 October 2015 on URL : http://www.bbc.com/news/world-middle-east-23849587, also accessed on 2 September 2017.
58. Neil Hauer, 'Russia's "Syrian People's Congress" in Sochi: Goals and Realities', Atlantic Council, 20 November 2017, available at http://www.atlanticcouncil.org/blogs/syriasource/russia-s-syrian-people-s-congress-in-sochi-goals-and-realities, accessed on 22 November 2017.
59. Hamidreza Azizi, 'Russia's Multilayered Syrian Diplomacy Worries Iran', *Al-Monitor*, 6 December 2017, available at http://www.al-monitor.com/pulse/originals/2017/12/iran-russia-sochi-summit-mistrust-astana-geneva-syria.html#ixzz51n7PmPxL, accessed on 6 December 2017.
60. Ibid.
61. Dmitri Trenin, 'Putin's Plan for Syria: How Russia Wants to End the War', *Foreign Affairs*, 13 December 2017, available at https://www.foreignaffairs.com/articles/syria/2017-12-13/putins-plan-syria?cid=int-rec&pgtype=art, accessed on 13 December 2017.
62. Ibid.
63. Brandon Friedman, 'Russia, Turkey, and Iran: Cooperation and Competition in Syria', *Tel Aviv Notes*, Vol. 11, No. 2, 30 January 2017, available at http://dayan.org/content/russia-turkey-and-iran-cooperation-and-competition-syria, accessed on 6 December 2017.
64. Ibid.
65. Ibid.
66. Trenin, 'Putin's Plan for Syria', n. 59.
67. Paul Bucala, 'Iran's New Way of War in Syria', A Report by the Critical Threats Project of the American Enterprise Institute and the Institute for the Study of War, 3 February 2017, available at https://www.criticalthreats.org/analysis/irans-new-way-of-war-in-syria, accessed on 2 September 2017.
68. Ibid.
69. Ibid.
70. For more, see Paul Bucala and Genevieve Casagrande, 'How Iran is Learning from Russia in Syria', A Report by the Critical Threats Project of the American Enterprise Institute and the Institute for the Study of War, 3 February 2017,

available at https://www.criticalthreats.org/analysis/irans-new-way-of-war-in-syria, accessed on 2 September 2017.
71. Ibid.
72. For more, see Ali Bakeer, 'Turkey, Iran, Russia: Trilateral Distrust in Syria', *Al-Monitor*, 20 November 2017, available at http://www.al-monitor.com/pulse/originals/2017/11/turkey-iran-russia-trilateral-distrust-in-syria.html#ixzz51n EC8bAk, accessed on 22 November 2017.
73. Friedman, 'Russia, Turkey, and Iran: Cooperation and Competition in Syria', n. 61.
74. Ibid.
75. For more, see Bakeer, 'Turkey, Iran, Russia: Trilateral Distrust in Syria', n. 70.
76. Azizi, 'Russia's Multilayered Syrian Diplomacy Worries Iran', n. 57.
77. For more, see Humud et al., 'Armed Conflict in Syria: Overview and U.S. Response', n. 54.
78. Ibid.
79. For more see Summary, Jeremy M. Sharp, Christopher M. Blanchard, "Armed Conflict in Syria: U.S. and International Response", 14 June 2013, Congressional Research Service, accessed on URL : https://www.everycrsreport.com/files/20130614_RL33487_15d32d142c4206ebb45fff142091d667a9b4b7f6.pdf,on September 2,2017.
80. Mark Mazzetti, Adam Goldman and Michael S. Schmidtaug "Behind the Sudden Death of a $1 Billion Secret C.I.A. War in Syria", 2 August 2017, The New York Times, accessed on https://www.nytimes.com/2017/08/02/world/middleeast/cia-syria-rebel-arm-train-trump.html, 2 September 2017.
81. For more see : Barnes, Julian E.; Entous, Adam (17 February 2015). "U.S. to Give Some Syria Rebels Ability to Call Airstrikes". The Wall Street Journal URL: https://www.wsj.com/articles/u-s-to-give-some-syria-rebels-ability-to-call-airstrikes-1424208053 ; and "House Grudgingly Approves Arms for Syrian Rebels". New York Post. Associated Press. 17 September 2014, URL: https://nypost.com/2014/09/17/house-grudgingly-approves-arms-for-syrian-rebels/. Both accessed on 2 September 2017.
82. See n. 78.
83. For more, see Humud et al., p. 31, 'Armed Conflict in Syria: Overview and U.S. Response', n. 54.
84. Ibid.
85. Ibid.
86. For more see "Trump Administration on ISIS, Al Qaeda", 15 August 2017, Wilson Centre, accessed on URL : https://www.wilsoncenter.org/article/trump-administration-isis-al-qaeda on 30 October 2017.
87. Ibid.
88. For more, see Humud et al., p. 31, 'Armed Conflict in Syria: Overview and U.S. Response', n. 54.
89. Ibid.
90. Ibid.
91. Jessica Lewis McFate, 'The ISIS Defense in Iraq and Syria', Middle East Security Report No. 27, May 2015, available at http://www.understandingwar.org/sites/default/files/ISIS%20Defense%20in%20Iraq%20and%20Syria%20—%20Standard.pdf, accessed on 1 September 2017

# 6

# Lebanon–Yemen Marathon
## Hezbollah Head and Houthi Legs

*Kishore Kumar Khera*

## Introduction

West Asia, with the Arab–Israel and Saudi–Iran rivalries as the prime drivers, at this juncture in history, is one of the most critical regions for world peace. Political, demographic and ideological stability in the region is at an ebb. Such an instability could easily expand both eastwards and westwards owing to presence of economically and politically vulnerable states. The character of conflict in this unstable zone since the 1970s has been transforming for two reasons: first, the realisation amongst the Arabs of their inability to defeat Israel in a conventional war; and second, playing out of the Iran–Saudi rivalry since the Islamic Revolution in 1979. This rivalry manifests itself through multiple means, with Yemen being the latest battleground. The Iran–Saudi conflict and the opposition to Israel in the region is not through direct force-on-force military engagements but in a hybrid form, through the use of proxies, use of military force by one against the proxies of another, propaganda, subversion, use of economic instruments, criminality, terrorism and so on, as contextualised in Chapter 2.

Over 2,000 kilometre (km) apart, Yemen and Lebanon typify the situation in West Asia. Both these coastal countries share land borders with two countries each. While Yemen, five times bigger than Lebanon, dominates one of the most active and strategic shipping lanes in the world, Lebanon, sandwiched between Syria and Israel, is located in one of the most volatile regions. Turmoil in these two countries and a similar type of conflict involving almost all facets of force application, barring nuclear, and the power play by multiple agencies, both internal and external, has enhanced the complexity of the operational environment. Transient and localised peace is interspersed with violence and bloodshed, along with attacks on various power tools and centres. Use of regular combatants, well-trained organised groups, mercenaries and civilians, especially technology experts, covers the human aspects. Technologically, a similar spread of crude and elementary munition to sophisticated, guided weapons is noticeable. 'Force-on-force' attritional confrontation is rare, and most of the engagements are indirect, discreet and often in the form of 'hit and hide'. Expansion of conflict domains to non-traditional areas, like electromagnetic spectrum, communications, cyber, information and psychological operations, demonstrates the ability and resolve of the warring factions. Use of terrorism, criminality and illegal economic resource generation has expanded the conflict arena.

State versus state military conflicts have primarily been direct force-on-force type of confrontations, with a small percentage of effort devoted to other elements. Earlier, owing to restricted access to high-end weapons, the non-state actors could only carry out operations with small arms low calibre weapons (SALW). However, in the last three decades, non-state actor empowerment has been the root cause for the transformation of conflict from SALW usage to high-end, long-range guided weapons and aerial platforms. The basic reason for this empowerment has been the expansion of military, technology and financial support to specific non-state actors by certain states, to achieve their political/ideological/economic objectives, bypassing force-on-force direct conflict. The world over, there has been an increase in and prolongation of 'grayzone' situations, that is, neither pure peacetime nor contingencies over territory, sovereignty and maritime economic interests.[1] With states sponsoring non-states actors, the infusion of military technology and finances has led to growing power and stature of non-state actors. This, in turn, has attracted high-end human resources, including technocrats, into its folds, hitherto restricted to lower end of socio-economic and educational strata.

This phenomenon is further accentuated by the availability of commercial off-the-shelf (COTS) technology and enhanced visibility. Revolution in communication technology, in general, and social networks and media tools, in particular, has resulted in greater reach and impact on the general populace. Communication technology, with little state control, has allowed a nearly unrestricted flow of information and financial resources. This has enabled non-state actors to have an international footprint. Coordination between various non-state actors based on their goals has become a reality, with the resultant deployment of high-end military technology to various parts of the globe. This has blurred the distinction between state and non-state, and also between conflict and peace, thus leading to hybrid wars. As Frank Hoffman has put it: 'Hybrid Wars can be waged by states or political groups, and incorporate a range of different modes of warfare including conventional capabilities, irregular tactics and formations, terrorist acts including indiscriminate violence and coercion, and criminal disorder.'

To understand the mechanics of hybrid warfare, two case studies are presented in this chapter: the Second Lebanon War (2006); and the ongoing conflict in Yemen. The following two sections deal with the kinetic and non-kinetic aspects of force application in Second Lebanon War and the current conflict in Yemen, respectively.

## Second Lebanon War

Lebanon, after the 1982 Israeli invasion, had to grapple with a long civil war. With a mixed population of approximately 59.7 per cent Muslims and 39 per cent Christians and the presence of foreign military forces, it remained a weak state.[2] To fill the power void, Hezbollah, a non-state actor, came into prominence, with ideological, financial, organisational and military support from Iran.[3] Primarily a Shiite outfit, it had a political and social welfare role and yet used, and still uses, violence as a tool, especially against Israel.[4] Societal composition and fragmented state apparatus allowed Hezbollah to develop its interests in southern Lebanon interplaying security and social roles.[5] The Lebanese internal situation changed with withdrawal of Israel from southern Lebanon in 2000 and the Syrian forces in 2005. But the state remained weak. In 2006, Hezbollah killed three and kidnapped two Israeli soldiers. Israel responded with 'Operation Change of Direction' that led to 34-day long Second Lebanon War aimed at decimating Hezbollah.[6] This is a classic case of a military engagement between a state and an external non-state actor. Now, the United Nations (UN), with one of the largest missions—United Nations Interim Force in Lebanon (UNIFIL)—has deployed over 12,400 peacekeepers to monitor cessation of hostilities and to support Lebanese Army in southern Lebanon.[7]

Israel Defense Forces (IDF) intent was to use superior firepower to pulverise Hezbollah and isolate their positions with tank manoeuvres. The main force application was with air power. Targets were Hezbollah strongholds, rocket launch systems, power, oil and infrastructure. The IDF flew over 19,000 sorties, dropping 20,000 bombs and firing 2,000 missiles from air, and nearly 125,000 artillery and heavy mortar shells were also expended, against almost 7,000 targets.[8]

With the active support of Iran, the military wing of Hezbollah was well trained and equipped akin to a regular army and was also well armed with guided munitions and unmanned aerial vehicles (UAVs). Use of radars and development of its optic fibre and cellular networks gave it a multifaceted capability. However, in the 2006 conflict, Hezbollah could not match the IDF in terms of resources or capability. Therefore, it resorted to guerrilla tactics in urban areas. 'Hit and hide' plan with ambushes and relocating to well-prepared defensive positions was the main ploy. This caused severe problems for the IDF.[9] Use of urban terrain, creation of defensive points, ability to use urban infrastructure for mobility and flexibility and well-planned and prepared weapons stocking areas allowed high-speed relocation for guerrilla warfare. Hezbollah defensive bunker systems were with electrical wiring, reinforced concrete fighting positions and enough water, food and ammunition to withstand a sustained siege.[10] Villages in southern Lebanon were also fortified to stall the IDF invasion and had over 500 arms storage sites.[11] Each village unit of Hezbollah was tasked to defend its location and delay IDF movement.[12] Training, skills, adaptability and leadership in each village cell were decisive for Hezbollah's performance. At every step, Hezbollah dictated the direction of conflict, and the IDF was only reacting in spite of overwhelming military superiority. Hezbollah's ability to manoeuvre tactically against the IDF, the autonomy given to its small units, the initiative taken by the small-unit leaders and the skill Hezbollah displayed with its weapons systems were a distinctive feature of Second Lebanon War.[13]

Hezbollah's main offensive inventory boasted of over 14,000 short- and medium-range rockets, of which 4,100 were fired. Its ability to strike deep inside Israel by rockets forced the evacuation of a number of Israelis. A good mix of offensive strategy was exhibited, with the use of rockets and armed UAVs, along with a defensive plan to slow down the progress of IDF. The effective use of the anti-tank guided missile (ATGM) systems, namely, RPG-29, AT-13 Metis and AT-14 Kornet, was key to the defensive battles. These ATGMs, with an effective range of 3–5 km against armour and battlefield targets, resulted in damaging 18 IDF tanks and killing approximately 52 personnel.[14] Induction and employment of three armed UAVs, probably Mirsad-1 or Ababil-3 (Swallow), with a range of 450km and payload of 50 kilogram (kg), by Hezbollah took the combat to another dimension, forcing the IDF to deploy aerial surveillance systems.[15] Adoption of available technology, from guided munitions against tanks and ships to UAVs, allowed Hezbollah to sustain itself in spite of limited conventional capability. Night operation facilitation by night vision goggles (NVG) and a missile attack on an Israeli ship took the conflict higher in the technology matrix.

While Hezbollah suffered four times the losses in manpower (estimated causalities for Hezbollah were about 600 and for IDF, 131) and its military capability was substantially reduced, it won the perception battle handsomely. The tactical employment of its kinetic force was augmented with a well-crafted information dissemination strategy. With well-defined yet distinct communication strategy for four different constituencies, namely, Shiite community, Lebanon population, the Arab world and international community, Hezbollah effectively utilised the art of psychological war. Production of television (TV) programmes in Hebrew and effective use of local and Palestinian photographers to highlight their condition increased popularity and reach of its TV channel, *Al-Manar*, in spite of a ban in certain countries.[16] This perception victory catapulted Hezbollah to a position of strength not only in Lebanon but also in the region. Israel, meanwhile, tried to counter this with an aggressive information and psychological warfare. It managed to break into *Al-Manar* transmission system and resorted to air dropping of leaflets.[17] Yet, it failed to counter Hezbollah's well-crafted media strategy. In fact, today, manipulating information on news channels/portals and use of digital communication to influence public opinion has gained special significance as an element of hybrid warfare.

High success rate and low attrition in the offensive missions by Hezbollah were indicative of a well-established intelligence network.[18] The electronic war was on simultaneously, with Hezbollah using communications intelligence (COMINT) and signals intelligence (SIGINT) to monitor IDF communications. Electronic intelligence (ELINT) and SIGINT equipment and training allowed Hezbollah to stay ahead in the game. Hezbollah's protection of transmission process of its TV channel during the height of Second Lebanon War showcased its ability to collate and interpret data into useful intelligence.[19]

'An army marches on its stomach' is a well-known adage bringing to fore the logistical planning to sustain dispersed and diffused operations as part of a hybrid conflict. Hezbollah realised the significance of a comprehensive and viable logistical plan to mitigate the impact of the overwhelmingly superior military power of IDF. The central theme of its logistical plan of dispersed storage was to deny Israeli Air Force an opportunity to achieve resource-neutralisation strike capability. Weapon supplied to Hezbollah, often through Syria or the sea route, were redistributed to various stocking points.[20] This dispersed and dynamic logistical plan was key to Hezbollah offensive and defensive operations. The plan's success was testified as Hezbollah fired over 250 rockets on the last day of the Second Lebanon War.

Iran and Syria were the main sources of Hezbollah finances. A number of large businesses in construction and real estate sectors in Lebanon were linked to Hezbollah.[21] Criminal activities, drug money and Bekaa Valley's poppy crop were used to bolster financial support.[22] Smuggling, kidnapping and extortion to raise, transfer and launder funds to achieve its goals was exemplified in June 2002, with the arrest of Muhammad and Chawki Hamud, in Charlotte, North Carolina.[23] Currently, though the Gulf Cooperation Council (GCC), Arab League and the European Union (EU) have declared the military wing of Hezbollah as a terrorist group, it continues to thrive economically through a combination legal and illegal activities.[24]

## Conflict in the Republic of Yemen

Yemen, post-unification in 1990, could not develop economically owing to poor policies and corruption, and thus remained one of the poorest West Asian country. With two-thirds of the population being Sunni, major power resided with them. In 2004, Hussein Badreddin al-Houthi, a Zaidi Shiite, organised a movement, 'Ansar Allah', opposing President Saleh's government. As a result, there were six armed conflicts from 2004 to 2010 in Yemen's northern province of Saada.[25] In 2011, with the general populace taking to the streets against government policies, the Houthis gave it the requisite momentum, forcing Saleh to quit. In February 2012, Vice President Hadi took over from President Saleh in a compromise to end the civil unrest. However, post a small pause, the conflict resumed between the government forces and the Houthis.

Yemen, as Bernard Haykel has pointed out, is a 'highly fragmented and divided country, with no national leadership that can unite a majority of the population around a vision or program for the future'.[26] While the GCC was successful in managing the transition from Saleh to Hadi, it has not yet succeeded in addressing the deeper political and economic malaise.[27] The failure of state institutions to adjudicate, arbitrate or mediate the social conflicts that polarise the polity and bring society to the tipping point is a core cause of civil war.[28] The Yemen crisis exemplifies this perfectly. Geographically, major parts of Yemen are under Hadi, with southern zone under Houthi control. Saleh, along with his loyalist section of Yemen Armed Forces, supported the Houthis who control the capital Sanaa. Saleh was killed in December 2017. This further complicated the power matrix. Besides these two major players, there are three other internal elements: Al Qaeda of Arabian Peninsula (AQAP), Al-Islah (Muslim Brotherhood of Yemen) and tribal leaders. Each of these five players is trying to dominate the others. The AQAP and local tribal groups are keen to exploit disarrayed security apparatus to dominate in small pockets around Lahij.[29] Fractured control and continuous conflict have forced millions of people to flee their homes and have killed or injured thousands.[30]

External forces in the fray are Saudi Arabia, Iran, Qatar and the United States (US). Operation Al-Hazm Storm, to defeat the Houthis and re-establish Hadi, was launched in 2015 by a Saudi Arabia-led coalition, including the United Arab Emirates (UAE), Kuwait, Bahrain, Jordan, Morocco, Sudan, Egypt and Pakistan. Saudi forces are coordinating with local tribal leaders and the UAE with proxy forces to enhance the effectiveness of military operations.[31] The US military has a dual role: one, attacking AQAP; and second, supporting Saudi Arabia. The US has supported Saudi-led coalition aircraft by providing aerial refuelling,[32] intelligence inputs, logistical support, along with $20 billion worth of military equipment to Saudi Arabia in 2016.[33] Iran supports the Houthis and Qatar is funding Al-Islah. Besides Syria and Iraq, the strategic rivalry between Saudi Arabia and Iran has intensified and become a major force behind the current fight in Yemen.[34] All forces involved in Yemen conflict are depicted in Figure 6.1.

In 2010, Yemeni Armed Forces had a strength of 66,700 and currently estimated strength of Yemen Army is about 20,000 owing to defections to support Saleh or for alternatives. Equipment destruction and attrition in the ongoing conflict has hit Yemen Navy and Yemen Air Force the hardest and these have practically ceased to exist.[35] Yemen Armed Forces are supported by the Saudi-led coalition against Houthi movement. The coalition commands over 500 fighter aircraft and requisite critical combat support elements, like ELINT, airborne early warning and control (AEW&C) system, Reconnaissance and Air to Air Refuelling. This is supported by a number of high-end surface-to-air weapon systems, like Patriot, for air defence of critical military and civil nodes. The surface forces include armour, artillery and infantry from the armed forces of the region, complemented by mercenaries hired from Latin America.[36]

**Figure 6.1: Pictorial Depiction of Warring Factions and External Support Elements in Yemen**

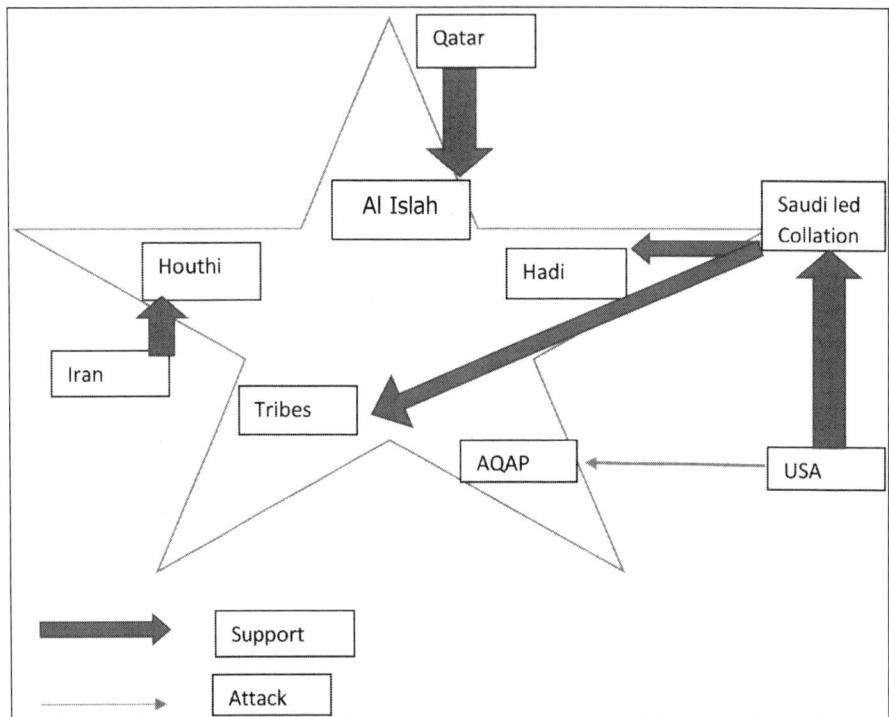

*Source*: Author.

On the other hand, the numerical strength of active members of Houthi movement engaged in armed conflict is estimated between 10,000–100,000.[37] However, assessing the area under active conflict and pace of operations, the actual numbers could be around 30,000–40,000 and the weapon operating elements between 10,000–15,000.[38] Currently, they have a limited stock of surface-to-air guided weapons (SAGWs), ATGMs, surface-to-surface missiles (SSMs), anti-ship missiles (ASMs) and UAVs.

Systematically and gradually, from 2012 onwards, the Houthis enlarged control of southern Yemen, including Dammaj. Houthis, in July 2014, supported by tribal leaders, in Amran, against Ahmars, gained territorial victories. In capital Sanaa, Houthis struck when demonstrations against corruption and removal of fuel subsidy, on 9 September 2014, turned violent, with police firing resulting in the death of eight protesters and an ambulance driver. A major Houthi attack on 16 September 2014 around a military camp north of Sanaa lasted four days, ending with their victory. This led to resignation by the prime minister and power balance tilting in favour of Houthis. To further consolidate, Al-Islah backer Sheik Hamid al-

Ahmar and his supporters were attacked. Capture of key military installations, the presidential palace in January 2015, resignation by Hadi and dissolution of parliament completed power shift to Houthi Governing Council. The Houthi plan was to contain AQAP and continue expansion to the oil-rich provinces of Marib and Al-Baydah.[39] Aden too came under Houthi control, albeit for a short while, as Yemen Army, reinforced by Saudi Arabia-led coalition air support, wrested it back on 15 July 2015. Immediately after this, the Houthis also lost control of Labuza army base in Lahij province north of Aden and the headquarters of the 117th Armoured Division in eastern Shabwa province. Coalition fire support, with rockets on Khor Maksar and air strikes on Anad air base, helped Yemen military regain lost ground. It is a known fact that Yemen military is continuously being equipped, trained and supported by Saudi Arabia and the UAE.[40] The UAE, the largest contributor of manpower for the coalition, deployed a brigade to defend Aden and has, till May 2017 lost 85 soldiers in this conflict.[41]

An analysis of the conflict dynamics in Yemen since 2004 indicates gradual upgradation in level, scale, tactics and technology, leading to battlespace expansion. At the beginning of the conflict, the occasional small arms fire on a Yemen Army patrol by Houthis, in contrast to use of combat aircraft to bomb the hostile locations by Yemen Air Force, indicated capability and technological gap between the two major players. With time, tactical engagements too shifted from 'hit and hide' to well-planned and coordinated ambushes using improvised explosive devices (IEDs). Organic growth of the organisation was boosted by military equipment and advice from Iran that reduced the tactical and technology gap and allowed the Houthis to gradually nibble territory and threaten military and civil targets with the use of long-range weapons. At present, the Houthis, though not as well trained and organised as Hezbollah, are resorting to same tactics as used by Hezbollah, that is, a mix of offensive and defensive, forcing Yemen Armed Forces along with the coalition to commit resources accordingly, resulting in the dispersal of forces to cover entire battlespace. Houthis, like Hezbollah, in spite of military capability inferiority, are dictating the terms of engagement through battlespace expansion—a typical hybrid attribute. The Houthis defensive positions are created in the urban centres, and the frequent offensive forays akin to guerrilla warfare have considerably slowed the progress of its opponents in spite of major capability differential. As is well known, success of hybrid tactics is based on creating small autonomous teams with offensive and defensive abilities and full-range exploitation of available equipment.

Expansion of spectrum of conflict is indicated by the Houthis firing the long-range Qaher-2 ballistic missile towards Jizan and Nazran in Saudi Arabia; attack on the US Navy destroyer, *USS Mason,* in the Red Sea in October 2016; and sniper attacks on a military base in Saudi Arabia and in Midi district. The attacks by the Houthis on the armoured vehicles of Saudi-led coalition forces—in Hawzan with an IED; in al-Mukha with ATGMs; in Harib al-Qaramish with artillery; and with rockets at Manfaz al-Khadra—indicates the span of munitions in use. Conflict in an open market in Marib and use of human shields by Houthi are signs of a shift of battlespace close to civil population and resultant civilian casualties. On 30 January 2017, a Saudi frigate was attacked near the Houthi-controlled port of Hodeidah by an unmanned remote-controlled boat laden with explosives. The attack killed one, and the frigate returned to port with minor damage.[42] In February 2017, a Yemen Coast Guard boat was destroyed near Al-Mokha by mines laid by the Houthis. Thus, technology infusion in the operational scenario is with UAVs, anti-tank and anti-ship missiles, as well as land and sea mines.[43] Technology will continue to help in greater diffusion of battlespace and enhance the hybridity of conflict.

The operational matrix indicates a robust command, control and intelligence network working for the Houthis wherein critical information about static and mobile systems in near-real time is available. To bolster its offensive capability, the Houthis plan to use Ababil series UAVs, fitted with high explosive warheads, to target radar of Patriot missile batteries to enhance the success rate their SSM attacks. Intelligence gathering and disseminating to the field operators has worked well in both Lebanon and Yemen. This is a key attribute for success of hybrid tactics using multiple means for single objective.

Logistics continue to play a pivotal role in Yemen. Logistical network for Houthi movement is inclusive of sea routes. However, the supply chain has severe blockages, resulting in a slow pace of operations and adequate time lapse between successive offensives. Throttling of supply routes, ban on oil exports and air attacks by Saudi-led coalition have forced the Houthis to remain defensive. However, Yemen Army, with better equipment and support from the coalition, has not yet been able to exploit available air superiority to regain lost ground owing to tough urban battles and logistical challenges.[44] Like other tenets of hybrid war, logistical plan needs to have multiple tentacles, with multiple routes, storages and sources, albeit all in small quantities.

When Houthi movement commenced in 2004, it was with fighters high on morale but low on military training. Protracted engagement for over

13 years has battle-hardened the group and the morale is still high owing to pull of greater power, including financial power with the capture of key areas and institutions like Central Bank of Yemen. In the initial phases, the weapon systems, mainly SALW, IEDs and mines, were procured from the black market or from the stocks of Yemen Army through corruption. However, the ongoing military engagement has seen the operation of specialist weapons and equipment from the Houthis. Operating SAGWs, ATGMs, SSMs and UAVs needs training and high skill levels. The Houthi movement is expected to have a limited number of specialist weapon operators, and these probably are trained by military professionals who were part of Yemen Armed Forces or with external help, probably from Iran and Hezbollah. Effective use of man-portable air defence systems (MANPADS) to shoot down aircraft of Yemen Air Force in 2014, and Saudi helicopter in 2017,[45] and the use of Yemen Air Force fighter aircraft to undertake air strikes on the presidential palace in Sanaa in March 2015 indicate defections from Yemen Air Force.[46] Use of sectarian card has allowed a flow of financial help and resultant induction of weapons for Houthis, followed by Houthis control of certain territories. Economic deprivation has nudged a large number of fence sitters to cross over to the Houthi movement against corrupt administration, leading to an exponential rise in its cadre and financial state. Houthis have remained focused on achieving a greater share in national power matrix, and therefore they have continued to build their combat capability. Seizure of military equipment and defections from Yemen Armed Forces have bolstered the already increasing military capability of the Houthis, which peaked in 2015–2016 with their ability to use fighter aircraft, UAVs, SSMs, SAGWs, tanks, infantry combat vehicles (ICVs) and unmanned explosive-laden boats. Presently, the Houthis have T-72, T-55, armoured infantry fighting vehicle (AIFV) and armoured personnel carrier (APC) to undertake surface operations, and a limited number of anti-aircraft artillery (AAA) batteries.[47]

Hybridity of conflict in Yemen gets complicated by simultaneous operations by different elements in the same geographical space, with distinct enemies and objectives. The US–AQAP battle adds another dimension. Commencing with a cadre strength of 200–300 in 2009, AQAP was estimated to have 1,500 personnel by 2015.[48] The AQAP, with two attempts to blow up aeroplanes with plastic explosives and detonator to be triggered by an alarm clock and a mobile phone, showed technological proficiency.[49] Houthi–Hadi conflict and the resultant chaos allowed AQAP to reorganise itself post multiple attacks by the US.[50] The AQAP, with suicide attacks on security forces in Mukalla and Aden, has retained relevance in conflict dynamics.[51] The US, to target AQAP, has carried out more than 80 strikes since 28 February 2017 in Yemen, as compared to about 30 strikes in 2016.[52] The US also carried out strikes on three coastal radars in Houthi-controlled coastal area in 2016.[53]

Zakat (almsgiving) was the main source of Houthi funding, along with donations from Iran and Hezbollah, prior to gaining control of Sanaa and the Central Bank of Yemen. To offset 53.7 per cent decline in revenue owing to the suspension of oil exports since 2015 and shrinking economic activity, the Houthis imposed a $20 war tax on civil servants and cut spending on budgetary items, like scholarships, that are unrelated to the war. Blackmarketing of fuel, the formation of a cartel of oil importers and control of non-banking systems have accelerated the rate of wealth transfer to the Houthis.[54] This has resulted in the sustenance of financial support for the conflict and high morale for the Houthi movement. Financial sourcing though legal, illegal and criminal means is evident in protracted hybrid war by non-state actors. For example, an investigation into Al Qaeda sleeper cells in Europe in the wake of the attack on World Trade Center in the US, on 11 September 2001, revealed the widespread use of legitimate businesses and employment by Al Qaeda operatives to derive income for supporting themselves and their activities.[55]

The next section covers the emerging contours, in which I delineate major common attributes of hybrid wars and geopolitical implications from these two case studies, as well as lessons for India.

## Emerging Contours

Conflict dynamics in Lebanon and Yemen show many similarities. Although there are no clear winners, the power equation continuously oscillates in various geographical and notional aspects, owing to failure of all players to define objective goals.[56] Ideological battle fought by both Hezbollah and Houthis has all the key elements of a hybrid war—kinetic and non-kinetic—but with different goals. The rise of Hezbollah was primarily against an external power, Israel, with strong religious flavour and little internal conflict with the state, whereas Houthi moment was initiated against the state for a greater share of power. Conceptual dissimilarities between the Houthis and Hezbollah end here. Hezbollah with a Shiite base and Houthi with a Zaidi base have Iran's support, albeit to different degrees based on ideology. Both these groups are focused on garnering state power. Operational philosophy and the scale of operations by the Houthis in Yemen are akin to what Hezbollah did in southern

Lebanon in its formative years. The prime motive of Hezbollah military wing is to protect Lebanon from Israeli attacks and therefore, it resorted to rocket attacks on the IDF and ambush of its patrols/posts while operating from southern Lebanon. While Hezbollah bore the brunt of IDF attacks in 2006, Houthis are facing a similar situation against Saudi-led coalition since March 2015, while engaged in operations to retain control of the territory from slipping back into the hands of Hadi-led government. Hezbollah, probably the strongest non-state actor in the region, has developed its military capabilities to oppose a strong IDF. The weak Yemen Armed Forces, supported by a Saudi-led coalition, have substantially lower military prowess as compared to the IDF and therefore, the nominal Houthi military capability has been able to retain its control over substantial territories. Over the years, Hezbollah has expanded its international footprint through ideological expansion, whereas the Houthis, with a socio-economic agenda, are restricted to Yemen. This represents the span of future hybrid threats: from being local to global, with varying capacities and capabilities, but invariably supported by state/states.

*Doctrinal Precepts: Compulsions of Hybrid Conflict*
Rules of engagement make the most distinct difference between the conflicting sides in this study. Hybrid tactics blur the boundaries between war and peace. The roles of aggressor and conflict party are deliberately obscured.[57] The conflicting sides in Lebanon and Yemen represent states and non-state actors. While Israel, Yemen and Saudi-led coalition represent the state and are bound by United Nations (UN) Charter on warfare, Hezbollah and the Houthis experience no such restrictions. Second distinction between the sides is capability exploitation: the non-state actors fully employ all their capabilities but state actors are forced to exercise restraint to contain collateral damage owing to diffused battlespace.

The air power was the most well-marked capability differential between the warring factions in Lebanon in 2006, and this is also true for the current Yemen situation. While Hezbollah was aware of its limitations in this aspect well before the commencement of the conflict, the Houthis realised it only after active participation by Saudi-led coalition. Hezbollah, therefore, developed a concept of command, control, communications and logistical support through dispersed and delegated subsets so as to deny Israel Air Force a centre of gravity for targeting. This also influenced Hezbollah's equipment profile and therefore, no manoeuvre capability with induction of tanks, was developed. The offensive element was dovetailed with defensive infrastructure and the plan was to exploit Israel's thin strategic depth by use of short- and medium-range rockets. To stall the movement of Israeli armoured formations, multiple small independent teams with ATGMs were developed and deployed. Hezbollah used medium-range rockets to expand the battlespace in 2006 and ATGMs to increase the depth of tactical engagements. Dispersal of Hezbollah cells in almost every village in southern Lebanon expanded the target area for IDF, and thus the reduction in force density.

*Operational Concepts: Execution of Hybrid Conflict*
A professional force like Israeli Air Force, with total air superiority and backed by robust intelligence, targeted over 7,000 sites in Lebanon and yet failed to comprehensively defeat a small Hezbollah force sans air support. Yemen conflict is also indicative of a similar outcome wherein all elements of Yemeni Armed Forces, including air strikes, were used to counter Houthi movement since 2004. In fact, the Saudi Arabia-led coalition has been undertaking air strikes against Houthi-held areas since 26 March 2015. The coalition started operations with an air campaign, that was later augmented with naval blockade and deployment of troops on the ground. The only deterrence that the Houthis could display against coalition air might was a limited number of MANPADS and AAA batteries. Post a pause for five months, the air strikes recommenced in August 2016, with focus on Saada, Sanaa, and Ta'az.[58] As a result of these air strikes, collaterals have formed a large number of 10,000 casualties since March 2015, in terms of women, children and medical staff of non-governmental organisations (NGOs).[59] Also, an air strike on the funeral procession of the father of the Yemeni interior minister on 8 October 2016 resulted in the killing of the local mayor and a number of tribal leaders.[60]

This reiterates that in a conflict against a dispersed and diffused opponent, the concepts of use of air power developed to tackle a conventional threat are of little use. In an urban/guerrilla warfare, air power utilisation needs to be transformed from an overtly offensive arm to a supporting, precise, intelligent and restrained component, so as to avoid collaterals and yet assist in achieving the laid-out objectives. Such an approach consumes an extraordinarily large amount of time and resources, backed by clearly thought-out strategy.

*Transforming Character of the Conflict*
States and non-state actors are resorting to methods of hybrid warfare involving the use of military means below the threshold of a conventional war to undermine a state in a covert manner. Weak states with non-

homogenous societies are particularly vulnerable. This approach combines various civilian and military means and instruments in a way that does not reveal their actual aggressive and offensive intentions until all pieces of the puzzle have been brought together.[61] While various explanations have been offered for hybrid warfare, in the white paper, 'Defence of Japan 2016', the term is used to mean 'aggression conducted by methods that are difficult to identify definitively as "armed attack" based on their outward appearance, involving a combination of non-military means, such as sabotage and information manipulation, and military means which are utilized covertly.'[62]

Looking at the components of Second Lebanon War and its after-effects and the ongoing Yemen conflict, it is evident that conflicts in future will have an application of all facets of power and the battlespace and battle timelines will be poorly defined. The conflicts will have political, ideological, social, economic and military sides and the battle of supremacy will be not only with military hardware but also with information and economy. Military employment will include urban conflict, irregular war, proxy war and guerrilla war simultaneously, and in the same space as direct 'force-on-force' engagement. Warfare is transforming from military-to-military direct engagement to a system-to-system engagement. Both in Lebanon and Yemen, ideological congregations have developed into politico-military entities with the use of kinetic and non-kinetic force to control the state and its socio-economic policies. Over a period of three decades, Hezbollah has matured and developed into a multifaceted organisation with an international footprint, in spite of ban by many nations. The Houthis are gradually stabilising as a force to control a part of Yemen independently and transforming from being a non-state actor to a state. Thus, it can be seen that 'ends justify the means' seems to be the underlying principle.

## Geopolitical Impact

Hezbollah, post-2006, has expanded its political, social, economic and military might in Lebanon and now practically controls the country. Passive role of the Lebanese Armed Forces during attacks by IDF in 2006 allowed Hezbollah to expand its influence in spite of heavy men and material loss and to garner a major share of perception battle as the main force defending Lebanon. Hezbollah has not only maintained its military wing against United Nations Security Council (UNSC) Resolution 1701 (2006) but also has expanded its strength and role.[63] Even with a small representation of 12 in 128-member Majlis an-Nuwwab (Lebanese Parliament), Hezbollah's power to veto allows it to control the nation without getting embroiled in the administrative process.[64] Its involvement in Syria and associated costs have diminished its stature. Based on how its leadership and supporters view its future role, it will have to strategise accordingly, but its military wing will continue to be the lynchpin which has converted a non-state actor to being a state actor sans state responsibilities. The situation experienced by Hezbollah in Lebanon and the lessons learned apparently are finding their way into Yemen. The ability of the Houthis to withstand military might of the coalition supported by the US, and containing AQAP and the local militia, is reminiscent of Hezbollah operational philosophy. The situation in Yemen is a typical multifaceted hybrid war, therefore a decisive termination of the military conflict is not foreseeable. It is likely to be a slow and long marathon race to peace using Houthi power and Hezbollah tactics, with support by Iran, against Saudi-led coalition supporting Hadi and Yemen Armed Forces. Fracture in GCC in June 2017, with a severance of ties with Qatar, has had an impact on the strength of the Saudi-led coalition force in Yemen. With over two years of conflict and a large number of collaterals, the US Special Operation Command is expected to increase its presence alongside Saudi Arabia and UAE troops in Yemen against the Houthis. As per UN Special Envoy to Yemen, Ismail Ould Cheikh Ahmed, 'A peace agreement, including a well-articulated security plan and the formation of an inclusive government, is the only way to end the war that has fuelled the development of terrorism in Yemen and the region.'[65]

The examination of events in West Asia indicates an intertwined power play between two major regional powers, Saudi Arabia and Iran. Yemeni affairs are further complicated by being the latest battleground for these two powers. While Saudi Arabia is backing Hadi, Iran is said to be supporting the Houthi rebellion. At the same time, Qatar is believed to be funding the Al-Islah party, which combines tribal leadership.[66] With the entry of Hezbollah on the side of the Assad regime in Syria and Hamas aligning itself with the Sunni axis, the sectarian divide in West Asia seems complete. The Hezbollah was till now a heroic Arab entity, an ally of Hamas, and supported by the Gulf sheikhdoms. By entering the Syrian conflict on a sectarian basis, it appears to have violated its own history. The influential Qatar-based cleric, Yusuf Qaradawi, who had earlier applauded it fulsomely when it had stood up to Israel in 2006, has now termed it the 'party of Satan'. He has also called on all Sunni Muslims with military training to march against Assad. Qaradawi was responding to the Hezbollah leader Hassan Nasrullah's pledge to support Assad till his final victory. This Saudi–Iranian confrontation has now also acquired a sharp sectarian character, with efforts across West Asia to mobilise a 'Sunni axis' to confront the 'Shiite crescent' led by the Islamic Republic of Iran.[67] The conflict is political, economic and tribal rather than a sectarian clash between Shiite and Sunnis.[68] Saudi Arabia had backed multiple individuals and factions in Yemen, including Al-Islah, till the uprisings of 2011. However, now Saudi Arabia has categorised and banned the Houthis and the Qatar-backed Muslim Brotherhood as terrorist organisations.[69] Persistent status quo on the definition of insurgency, freedom struggles and terrorism among various states restrains the process to bring peace. Viewing all events as a zero-sum game limits the perceptions of this issue.[70]

An acute sense of vulnerability is reflected in the nature of conflicts in West Asia, and vertical cleavages are deepening on sectarian basis with the main cause of socio-economic disparity taking a secondary role. As per Timo Kivimaki, protection wars have become the main course of violence in the world, occasionally contributing over 50 per cent of total fatalities.[71] In the propensity to garner strategic significance, Saudi Arabia and Iran have taken the conflict in Yemen to a high level on the technological matrix and expanded the civil disturbance to a hybrid war. With multiple entities aiming for a favourable end state, the conflict, though of low to moderate intensity, will continue for a long time owing to overlapping objectives. This prognosis is supported by a dynamic simulation approach for evaluating the scope and intensity of the conflict trap[72] and is reaffirmed by poor record of West Asia socio-political engineering through the use of coercive force.[73] The deepening of the sectarian divide in Lebanon, Hezbollah's focus on the Islamic State of Iraq and Syria (ISIS) and sustenance of AQAP are ominous signs of a conflict trap that West Asia will find hard to come out from. Qatar crisis in GCC has added another dimension to this vulnerable region. And maybe at the end of it all, proxy wars and hybrid wars are facilitated whenever the states, without a visionary leader, are a kaleidoscope of religions, regions and traditions and just ready to be exploited. This makes it difficult to identify agencies to be engaged for a lasting solution.

## *Lessons for India*

Open, pluralistic and democratic societies offer a number of potential targets and are therefore particularly vulnerable to hybrid activities.[74] Hybrid war, with multiple tentacles and poorly defined beginning and end, is a reality, with its perpetrators including states and non-state actors acting from within or outside the state. Preparations as a nation need to be made to thwart a combination of cyberattacks, propaganda, economic pressure, political destabilisation with irregular and covert elements, subversion and regular armed forces. Therefore, the armed forces must be ready to face a spatially focused threat posed by military forces below or above the threshold of open warfare, at short notice, as part of a hybrid strategy characterised by the orchestrated use of military and non-military means across the full range of the threat spectrum.

For kinetic aspects, development of quick processes, accurate offensive tools and robust defensive capabilities are necessary. Ability to undertake high-intensity operations in urban areas and yet restrict the collateral damage is essential. Battlespace in such a scenario is often fluid and multifaceted and capability for real-time surveillance of entire expanse is a prerequisite for operational success. In non-kinetic arenas, multiple identities of Indians based on region, religion, language and caste, and divide based on socio-economic and political paradigm, become a significant factor. This coupled with a rapid expansion of communication tools leaves low reaction time for state apparatus to contain the aggression. While the threat needs to be tackled in multiple dimensions, the best methodology to tackle it is at the beginning, owing to its low initial intensity, and to firewall areas and sectors for lateral propagation. Entire capability needs to be applied to thwart its spread. Education and governance transparency are two most significant tools for the purpose on the domestic front. Internationally, with Indian diaspora spread in various regions of the world, the situation needs to be assessed critically and requisite advisory issued to avoid endangering them prior to and during panic evacuations.

## Conclusion

The Second Lebanon War in 2006 and the ongoing conflict in Yemen indicate the hybrid nature of warfare, in which a political/socio-economic narrative supported by armed elements utilising flexible organisational structure and technology can withstand a militarily more powerful adversary. Hezbollah, with a small core group of 1,000+, adapted and prepared the battlespace to their strengths. A delegation of powers to each cell and their ability to augment the cells with local villagers and basic weaponry skills eliminated the need for a long command, control and logistical chain. This, in turn, made it difficult for IDF to identify critical targets and hit them. A similar model is being followed by Houthis in Yemen, albeit against a less determined and capable enemy. While Hezbollah has practically garnered control of Lebanon and is deeply embroiled in Syria, Houthis are yet to reach a state of equilibrium in Yemen. Both conflicts, separated by a decade, indicate the transformation of nature, scope and basis of warfare.

Threats to the state and society are not always easy to predict. The opportunities offered by globalisation, new technologies and the digital age are redefining the conduct of conflict. This holds an important lesson about the type of forthcoming threats for states and societies and to be prepared accordingly.

## NOTES

1. White Paper on Defence of Japan, 2016, available on http://www.mod.go.jp/e/publ/w_paper/pdf/2016/DOJ2016_Digest_part1_web.pdf, accessed on 31 August 2017.
2. 'The Lebanese Demographic Reality', Lebanese Information Center, Lebanon, 14 January 2013, p. 1.
3. Daniel Byman, 'Is Hezbollah Less Dangerous to the United States?', 18 October 2016, available at https://www.brookings.edu/blog/markaz/2016/10/18/is-hezbollah-less-dangerous-to-the-united-states, accessed on 18 May 2017.
4. David Thaler, 'The Middle East: The Cradle of the Muslim World', in Angel M. Rabasa, Cheryl Benard, Peter Chalk, C. Christine Fair, Theodore Karasik, Rollie Lal, Ian Lesser and David Thaler, *The Muslim World after 9/11*, Santa Monica, CA: Rand Corporation, 2004, p. 121.
5. Theodore Karasik and Cheryl Benard, 'Muslim Diasporas and Networks', in Rabasa et al., *The Muslim World after 9/11*, n. 4, p. 459, indicate that Hezbollah set up a network of humanitarian facilities and medical care in Lebanon's collapsed health care system, predominantly in Shia areas and southern Lebanon. In addition to providing general health care, the medical centres had delivery wings and care for newborns, laboratories, x-ray departments that specialised in care for women and children, a vaccination section that also provided general health advice, a blood bank and dispensaries. Bakeries provided nutritious food. Funding for Hezbollah's health sector came from Iran and the Lebanese diaspora, including communities in West Africa, notably Nigeria. Other Iranian non-governmental organisations (NGOs) contributing to Hezbollah's health network included the Imam's Relief Committee, Foundation of Martyrs, Construction Crusade, Housing Foundation and the Volunteer Women's Community Health Workers' Organization.
6. Martin N. Murphy, *Small Boats, Weak States, Dirty Money*, London: Hurst Publishers, 2009, p. 293.
7. 'UNIFIL Mandate', available at http://www.un.org/en/peacekeeping/missions/unifil/, accessed on 23 May 2017.
8. Frank Hoffman, *Conflict in the 21st Century: The Rise of Hybrid Wars*, Virginia, US: Potomac Institute for Policy Studies, 2007, p. 39.
9. Andrew Exum, 'Hizballah at War: A Military Assessment', Policy Focus No. 63, Washington Institute for Near East Policy, Washington, DC, December 2006, pp. 9–11.
10. Ibid., p. 3.

11. Ibid., p. 4.
12. Ibid., p. 10.
13. Ibid., p. 13.
14. Ibid., pp. 9–14.
15. Ibid., p. 5.
16. Thomas Rid and Marc Hecker, *War 2.0*, New Delhi: Pentagon Press, 2010, pp. 141–161.
17. Hoffman, *Conflict in the 21st Century*, n. 8.
18. Karasik and Benard, 'Muslim Diasporas and Networks', n. 5, p. 446.
19. Rid and Hecker, *War 2.0*, n. 16.
20. Joseph Daher, 'Reassessing Hizbollah Socioeconomic Policies in Lebanon', *The Middle East Journal*, Vol. 70, No. 3, 2016, p. 402. Also see Murphy, *Small Boats, Weak States, Dirty Money*, n. 6, p. 298.
21. Daher, 'Reassessing Hizbollah Socioeconomic Policies in Lebanon', n. 21, p. 416.
22. Karasik and Benard, 'Muslim Diasporas and Networks', n. 5, p. 468.
23. Ibid., p. 468. Also see Zoe Hu, 'The History of Hezbollah, from Israel to Syria', *Lebanon News*, December 20, 2016 available on http://www.aljazeera.com/news/2016/10/history-hezbollah-israel-syria-161031053924273.html accessed on 8 February 2018
24. 'Arab League Labels Hezbollah a "Terrorist" Group', *Al Jazeera*, 12 March 2016, available at http://www.aljazeera.com/news/2016/03/arab-league-labels-hezbollah-terrorist-group-160311173735737.html, accessed on 18 May 2017.
25. Iona Craig, 'What the Houthi Takeover of Sanaa Reveals about Yemen's Politics', *Al Jazeera America*, September 25, 2014, available at http://america.aljazeera.com/articles/2014/9/25/houthi-yemen-takeover.html, accessed on 8 May 2017.
26. Bernard Haykel, 'Yemen's Unsteady Transition', *Al-Sharq Al-Awsat*, 16 April 2013, available at http://www.aawsat.net/2013/04/article55298896, accessed on 8 May 2017.
27. Farea Al Muslimi, 'Yemen's Southern Movement Seeks Real Change from Sanaa', *The National*, 16 March 2013, p. 11.
28. Nazih Richani, 'The Political Economy and Complex Interdependency of the War System in Syria', *Civil Wars*, Vol. 18, No. 1, 2016, p. 48.
29. 'Chronology', *The Middle East Journal*, Vol. 71, No. 1, 2017, p. 145.
30. United Nations Office for the Coordination of Humanitarian Affairs, 'Yemen: We must Act Now to Prevent a Humanitarian Catastrophe', 24 April 2017, available at https://www.unocha.org/country/node/269703, accessed on 5 May 2017.
31. International Institute for Strategic Studies (IISS), *Military Balance 2017*, London: IISS, 2017, pp. 354–355.
32. Mathew Rosenberg and Mark Mazzetti, 'U.S. Ship Off Yemen Fires Missiles at Houthi Rebel Sites', *The New York Times*, October 12, 2016, available at https://www.nytimes.com/2016/10/13/world/middleeast/yemen-rebels-missile-warship.html?_r=0, accessed on 15 June 2017.
33. 'Chronology', *The Middle East Journal*, n. 30, p. 133.
34. Gawdat Bahgat, 'Lower for Longer: Saudi Arabia Adjusts to the New Oil Era', *Middle East Policy*, Vol. XXIII, No. 3, 2016, p. 47.
35. IISS, *Military Balance 2017*, n. 32, pp. 412–413.
36. Emily Hager and Mark Mazzetti, 'Emirates Secretly Sends Colombian

Mercenaries to Yemen Fight', *The New York Times*, 25 November 2015 available on https://www.nytimes.com/2015/11/26/world/middleeast/emirates-secretly-sends-colombian-mercenaries-to-fight-in-yemen.html and accessed on 8 February 2018

37. Hakim Almasmari, *Yemen Post*, available at http://yemenpost.net/Detail123456789.aspx?ID=3&SubID=1723&MainCat=4, accessed on 22 May 2017.
38. IISS, *Military Balance 2016*, London: IISS, 2016, p. 360.
39. Craig, 'What the Houthi Takeover of Sanaa Reveals about Yemen's Politics', n. 26.
40. Noah Browning, 'Anti-Houthi Forces Advance in Yemen amid Heavy Arab Air Strikes', 18 July 2015, available at http://www.reuters.com/article/us-yemen-security-idUSKCN0PS08T20150718, accessed on 5 May 2017.
41. 'Chronology', *The Middle East Journal*, Vol. 70, No. 1, 2016, p. 144; and Chris Tomson, 'UAE Soldier Killed as Houthi Forces Assault Saudi-held Port City in Western Yemen', 3 May 2017, available at https://www.almasdarnews.com/article/uae-soldier-killed-houthi-forces-assault-saudi-held-port-city-western-yemen/, accessed on 5 May 2017.
42. Thomas Gibbons-Neff, 'Houthi Forces Appear to be Using Iranian-made Drones to Ram Saudi Air Defenses in Yemen, Report Says', *The Washington Post*, 22 March 2017, available at https://www.washingtonpost.com/news/checkpoint/wp/2017/03/22/houthi-forces-appear-to-be-using-iranian-made-drones-to-ram-saudi-air-defenses-in-yemen-report-says/?utm_term=.76806058af7e Thoam= accessed, accessed on 5 May 2017.
43. 'US Admiral Says Power of Yemeni Houthi Fighters Growing', *Press TV*, 29 April 2017, available at http://www.presstv.ir/Detail/2017/04/29/519873/Yemen-Houthis-Vice-Admiral-Kevin-Donegan, accessed on 5 May 2017.
44. As per reports, Brigadier General Ahmed Asseri, from the Arab coalition, said, 'We don't lack information or evidence that the Iranians, by various means, are smuggling weapons into the area. We observe that the Kornet anti-tank weapon is on the ground, whereas before it wasn't in the arsenal of the Yemeni army or of the Houthis. It came later.' Weapons from Iran arrive either directly or via Somalia through sea route, getting transhipped to small fishing boats to escape detection in 2,700 km of coastline around Yemen. On 3 May 2017, Conflict Armament Research (CAR) reported similarities between Houthi Qasef-1 drones and drone engine with Iran Aircraft Manufacturing Industrial Company's product, Ababil-T. Six of these drones were captured in October 2016 on a known Iranian smuggling route through Oman, while another was found after an attack by Houthi forces near Aden in April 2017. In addition to specialist weapons like ASM and man-portable air defence systems (MANPADS), Iranian role is assessed in tactical military and logistical advice. See Jonathan Saul, Parisa Hafezi and Michael Georgy, 'Iran Steps up Support for Houthis in Yemen's War', 21 March 2017, available at http://www.reuters.com/article/us-yemen-iran-houthis-idUSKBN16S22R, accessed on 5 May 2017.
45. Leith Fadel, 'Houthi Shoot Down Saudi Helicopter', 25 January 2017, available at https://www.almasdarnews.com/article/houthi-forces-shoot-saudi-chopper-yemen/, accessed on 22 May 2017.
46. 'Yemen Air Force', available at http://www.globalsecurity.org/military/world/yemen/airforce.htm, accessed on 22 May 2017.
47. IISS, *Military Balance 2016*, n. 39, p. 360.

48. Rezaei Farhad, 'Shopping for Armageddon: Islamist Groups and Nuclear Terror', *Middle East Policy*, Vol. XXIII, No. 3, 2016, p. 123.
49. Ibid., p. 124.
50. Ibid.
51. 'Chronology', *The Middle East Journal*, n. 30, p. 144.
52. Cheryl Pellerin, 'Pentagon Spokesman Updates Iraq, Syria, Yemen Operations', US Central Command, 24 April 2017, available at http://www.centcom.mil/MEDIA/NEWS-ARTICLES/News-Article-View/Article/1162256/pentagon-spokesman-updates-iraq-syria-yemen-operations, accessed on 15 June 2017; and Bethan Mckernan, Al-Qaeda claims it is 'fighting alongside' US-backed coalition forces in Yemen, Independent, 2 May 2017 available at http://www.independent.co.uk/news/world/middle-east/al-qaeda-group-leader-claims-fighting-alongside-us-backed-coalition-forces-yemen-a7713321.html, accessed on 5 May 2017.
53. 'Chronology', *The Middle East Journal*, n. 30, p. 146.
54. Amal Nasser, 'How Long can the Houthis Hold on to Yemen?', *Al-Monitor*, 13 September 2016, available at http://www.al-monitor.com/pulse/originals/2016/09/yemen-war-funding-sources-houthis-saudi-arabia.html, accessed on 5 May 2017.
55. Karasik and Benard, 'Muslim Diasporas and Networks', n. 5, p. 464.
56. Zui Bar'el, 'Tahrir Square, From Place to Space: The Geography of Representation', *The Middle East Journal*, Vol. 71, No. 1, 2017, p. 22.
57. 'White Paper on German Security Policy and the Future of Bundeswehr', 2016, available at https://www.bmvg.de/resource/resource/.../2016%20White%20Paper.pdf, accessed on 14 July 2017.
58. 'Houthi Forces Deliver Another Blow to Saudi-led Forces in Yemen', 4 May 2017, available at https://southfront.org/houthi-forces-deliver-another-blow-to-saudi-led-forces-in-yemen, accessed on 5 May 2017.
59. 'Chronology', *The Middle East Journal*, n. 30, p. 145.
60. Ibid., p 146.
61. 'White Paper on German Security Policy and the Future of Bundeswehr', n. 59.
62. White Paper on Defence of Japan, 2016, n. 1.
63. UNSC Resolution 1701 of 2006 calls for: 'full implementation of the relevant provisions of the Taif Accords, and of resolutions 1559 (2004) and 1680 (2006), that require the disarmament of all armed groups in Lebanon, so that, pursuant to the Lebanese cabinet decision of 27 July 2006, there will be no weapons or authority in Lebanon other than that of the Lebanese State;'
64. Tim Butcher, Hizbollah's Lebanon veto power boosts Iran's Middle East influence, *The Telegraph*, 21 May 2008 accessed on 8 February 2018
65. 'The Death Toll in Yemen Conflict Passes 10,000', 17 January 2017, available at http://www.aljazeera.com/news/2017/01/death-toll-yemen-conflict-passes-10000-170117040849576.html, accessed on 5 May 2017.
66. Michael Peel, 'Rivals make Play for Power in Yemen', *Financial Times*, 15 April 2013, p. 6.
67. Talmiz Ahmad, *The Islamist Challenge in West Asia: Doctrinal and Political Competitions after the Arab Spring*, New Delhi: Pentagon Press, 2013, p. viii.
68. Craig, 'What the Houthi Takeover of Sanaa Reveals about Yemen's Politics', n. 26.
69. Ibid.

70. C. Christine Fair, 'Iran: What Future for the Islamic State?', in Rabasa et al., *The Muslim World after 9/11*, n. 4, p. 234.
71. Timo Kivimaki, First Do No Harm: Do Air Raids Protect Civilians?, *Middle East Policy*, Vol. XXII, No. 4, 2016 available on http://www.mepc.org/journal/first-do-no-harm-do-air-raids-protect-civilians accessed on 8 February 2018.
72. Havard Hegre, Havard Mokleiv Nygard and Ranveig Flaten Raeder, 'Evaluating the Scope and Intensity of the Conflict Trap', *Journal of Peace Research*, Vol. 54, No. 2, 2017, p. 259.
73. Merom, 'The Logic and Illogic of an Israeli Unilateral Preventive Strike on Iran', n. 20, p. 107.
74. 'White Paper on German Security Policy and the Future of Bundeswehr', n. 59.

# 7

# Israel and the Challenges of Hybrid Warfare

*S. Samuel C. Rajiv*

## Introduction

The 2006 Israel–Hezbollah War has been termed as 'the poster child of hybrid warfare'.[1] Hezbollah, a non-state armed group (NSAG), effectively combined the usage of sophisticated conventional weapons like anti-tank missiles and unmanned aerial vehicles (UAVs) with irregular warfare tactics and strategic communication efforts to put up a strong fight against a militarily stronger nation-state. The 2006 war most closely resembles Frank Hoffman's definition of hybrid warfare as 'the blend of lethality of state conflict with the fanatical and protracted fervor of irregular war'.[2]

Theoreticians of the history of warfare, however, argue that hybrid warfare is not a new phenomenon but predates the 2006 Lebanon War. Elements of its usage can be found as far back as the Peloponnesian wars as well as in such conflicts such as the American Revolution, Napoleonic wars (Spain and Russia), Communist Revolution in China and the Indochina wars, among others. All of these 'compound wars' saw a conventionally superior force fight a mix of guerilla and conventional forces.[3] They further note that these kinds of conflicts are likely to become the norm of future warfare. At the same time, while nation-states have faced hybrid threats in increasing order, they themselves have used hybrid elements at their disposal to out step their opponents.

The aim of this chapter is to discern the hybrid characteristics of conflicts that Israel has been involved in, both as a proponent who has employed such tactics as well as in facing down opponents who have utilised hybrid threats against it. For purposes of this chapter, three conflict situations are examined. These include the 2006 Israel–Hezbollah War; the three Gaza conflicts post-2005 disengagement; and the Iranian nuclear challenge. These conflict situations correspond to the 'rainbow of conflict' that Israel had to grapple with, ranging from low intensity conflict (LIC) in Gaza, to high intensity conflict (HIC) in Lebanon, to conflict with states without a common border (Iran).[4]

In tune with the philosophy of this book, the chapter seeks to distil Israeli use (as well as countering the use) of kinetic and non-kinetic elements in the pursuit of political goals in these three case studies.[5] First, however, a brief overview of the Israeli strategic situation and use of force/military doctrine vis-à-vis state-centric threats, non-state threats and weapons of mass destruction (WMD) threats is explicated. The section seeks to place in context Israel's responses to the myriad security challenges it had to face since its founding.

Even prior to its founding as a modern nation-state in 1948, it is pertinent to flag the irregular warfare tactics adopted by Jewish resistance organisations from the Roman times (the 'zealots') to those like the Irgun and Stern Gang against the British imperial power during the period of the British Mandate for Palestine. The bombing of the King David Hotel in July 1946 by the Irgun, which killed 90 people, is illustrative of what has been termed 'Jewish terror'. The Jewish paramilitary organisation 'Haganah', in fact, became the core of the Israel Defense Forces (IDF) after the nation-state was founded. The depopulation of Arab villages by use of force or threat of use of force by Jewish organisations, and the counter-campaign of terror by the Arab population, not only continues to be a subject of much critical contention but is also illustrative of kinetic and non-kinetic measures which were employed by both sides of the divide even prior to the establishment of the Jewish state.

## Overview of Israel's Strategic Situation and Use of Force/Military Doctrine

### State-centric Threats

Since its founding in 1948, Israel has fought six wars with its Arab neighbours. These include the combined Arab onslaught in the immediate aftermath of its founding in 1948, the 1956 Sinai Campaign, the Six-Day War of 1967, the 1969–1970 War of Attrition (with Egypt primarily), the 1973 Yom Kippur War and the 1982 Lebanon War. Analysts note that while the 1948 War of Independence along with the 1969–1970 War of Attrition and the 1973 Yom Kippur War were 'defensive' wars (forced upon Israel by its Arab enemies), the 1956 Sinai Campaign was a 'preventive' war (undertaken in the aftermath of the 1955 Soviet–Czech arms sales to Egypt which Israel believed would disproportionately tilt the balance of power in Egypt's favour, among other reasons) and the 1967 war was a 'pre-emptive' war (undertaken to preclude its enemies from carrying out their intended military attacks).[6]

The 1948 War of Independence brought into sharp focus the Arab antipathy to its existence as an independent Jewish nation-state in their midst. Given its geographical limitations and the limited strategic depth such geography conferred on it, Israel's military leadership was acutely aware of the need to transfer the fighting onto enemy territory as quickly as possible when hostilities broke out. This was on account of the fact that defeat in any of the wars Israel fought with the Arab states could have entailed the end of the state. Former Defence Minister Yigal Allon's statement that 'either you win the war or you will be driven into Mediterranean' captures the sense of the Israeli security predicament.[7]

Israel, therefore, privileged an offensive preventive or pre-emptive tactical posture in its unwritten military doctrine to ensure strategic defence. The IDF was mainly structured for the offensive, at the expense of its defensive capabilities.[8] Former Vice Chief of the IDF Israel Tal's formulation that the IDF is the '"Israel Defense Force" by appellation but the "Israel Offense Forces" in substance' best captures the nature of its military profile prior to 1967.[9] Offensive posture was seen as the best strategy to make up for Israel's limited strategic depth, limited manpower resources as against its Arab enemies and lack of alliance partners, among other factors.[10] The principal tactics involved in such an offensive warfighting doctrine were lightning armoured thrusts into enemy territory along with the use of air power.

When Israel's sense of security improved after the 1967 war—as a result of the concomitant increase in strategic depth after it captured Sinai, Golan Heights, West Bank and East Jerusalem—it enhanced its strategic defensive posture. This posture was continued even after the deterrence setback it suffered during the 1973 Yom Kippur War, when it had to face down the combined surprise onslaught by Syria and Egypt. Among reasons analysts attribute to the continuation of such a posture include: dependence on the United States (US), which to some extent limited Israel's freedom of action; a defensive posture being more politically expedient domestically; the need to build up the IDF's capabilities relating to firepower and mobility; and a renewed focus on territorial defence, an issue which was neglected during the 1967–1973 period.[11]

After the coming to power of the hawkish Menachem Begin in the late 1970s and the political ascendance of Ariel Sharon thereafter, Israel again began to privilege offensive pre-emptive actions against purported threats. These were reflected, among others, in Israel's raid on Osiraq in June 1981 and the initiation of the 1982 Lebanon War to drive out the Palestine Liberation Organization (PLO) from Beirut. The 1991 Gulf War, meanwhile, brought to Israel's stark attention the lack of deterrence against a state with which it did not share a border and which had medium-range ballistic missiles at its disposal. Israel's dominant military doctrine based on the armoured offensive manouvre—termed the 'cult of the offensive' by critics—proved irrelevant in the face of the Iraqi missile onslaught.[12] The possibility of incurring large-scale civilian casualties as a result of the Iraqi missile strikes targeted at its 'rear' consumed the Israeli political leadership during the critical days of the Gulf War. Prime Minister Yitzhak Shamir's government, however, exercised remarkable restraint and did not retaliate against these strikes.

The use of air power, 'techno-centric warfare' and 'effects-based operations' (involving massive air strikes) in the Gulf War, according to analysts, impinged negatively on Israeli thinking and was most reflected during the 2006 Lebanon War (see the later sections for an analysis of the 2006 war). The overthrow of Saddam Hussein in 2003 and the turmoil in Syria since 2011 also diminished the lethality of the two primary state-centric threats to Israeli security. The three Gaza conflicts since disengagement in 2005, as well as the 2006 Lebanon War, further reinforced the hybrid nature of threats that Israel had to grapple with, which combined elements of LIC and HIC.

Apart from the six major inter-state wars, Israel's enemies also indulged in regular and irregular fighting in periods like those between 1953–1956 and 1967–1970 (Egypt; and Jordan in 1953–1954). Israel's preferred choice in such circumstances was limited military retaliation, though it tried to stimulate ambitious policy options like regime change when it indulged in 'deep-penetration' bombing campaign in the vicinity of Cairo in 1970 with the aim of striking at the heart of the Nasserite regime. Analysts however note that Israel was unsuccessful as the target sets were inappropriate (training facilities were targeted instead of sensitive leadership targets), bombing raids were far too limited, no clear threat was communicated and Nasser subsequently survived with Soviet military support.[13]

*Non-state Threats*

The Israel Security Agency (*aka* 'Shin Bet' after its Hebrew acronym) currently lists 16 terror organisations that pose a threat to Israel. These include the following: Al Aqsa Martyrs Brigades, Al Qaeda, Army of Islam, Democratic Front for the Liberation of Palestine, Fatah Al Islam, Hamas, Hezbollah, Popular Front for the Liberation of Palestine or PFLP (Ahmed Jibril faction), People's Army, Salah al-Din Brigades, the Abu Ali Mustafa Brigades, the Izz al-Din al-Qassam Brigades (military wing of Hamas), the Jerusalem Brigades, Palestinian Islamic Jihad, PFLP and the Popular Resistance Committees (PRC).[14]

Most of these organisations owe their existence to the long-running Israel–Palestine dispute, which shows no signs of abatement. The Israeli–Palestinian conflict was described as a 'protracted social conflict' even as far back as 39 years ago.[15] The intifada (uprising) during the 1980s and 2000s presented a unique challenge to the IDF as it was 'neither a guerilla war nor a terror campaign' but a combination of both that sought to undermine Israeli security and administrative control over the territories.[16] Initially, it was seen as a law and order problem, but the IDF gradually became more involved after the severity of the threat increased. The Second Intifada moreover corresponded with rise in suicide bombings and rising Israeli fatalities.

According to the Israeli government, 1,328 civilians have been killed as a result of Palestinian terrorism and violence since 2000 till July 2017;[17] and nearly 40 per cent of the fatalities were due to suicide bombings. It is pertinent to note that during the two decades prior to 2000, about 510 Israelis lost their lives as a result of Palestinian terrorism.[18] Israel has resorted to both offensive and defensive measures to tackle such treats. Targeted killings of terror masterminds and heads of organizations has been one such prominent offensive tactic. Some of the key terrorist leaders killed by Israel include the Hamas 'bomb maker', Yahya Ayyash (January 1996); Hamas's founder, Sheik Ahmed Yassin (March 2004); Hamas's co-founder, Abdul Aziz Rantissi (April 2004); PRC commander, Zuhair al-Qaissi (March 2012); and Hamas military chief, Ahmed al-Jabari (November 2012), among others.

These measures have been the subject of much scrutiny, though Israel swears by their efficacy. Analysts note that offensive actions against terrorists and their infrastructure, which while not solving terror problems, cause 'short- and medium-term damage' to the terror organisation and their leaders, which constrict their ability to mount subsequent attacks. However, the other side of the argument has been that such offensive actions lead to further radicalisation and 'revenge' attacks leading to the death of more Israelis.[19]

The building of the defensive perimeter separating the West Bank and the Gaza Strip is an example of one prominent defensive measure. The Israeli government contends that the barrier has been an effective impediment to terror strikes, especially suicide bombing attacks. Analysts note that the efficacy of such defensive measures has, in fact, resulted in a change of Palestinian terror tactics, with an increased focus on rocket attacks in order to overcome the limitations imposed by the security barrier. Israel has undertaken major military strikes after the 2005 disengagement in order to counter such rocket attacks against civilian populations (see later sections for analysis of the Gaza conflicts). Apart from such offensive and defensive measures, the IDF has also resorted to use of innovative tactics in fighting urban terror. The IDF UAVs, for instance, were used in the Jenin refugee camp in Operation Defensive Shield in 2002. The IDF, in fact, has the distinction of operational use of UAVs for over four decades. Given that platforms like UAVs have become an essential element of what has been termed 'Fourth Generation Warfare', Israel's role in the development and use of such platforms has been pivotal.

The 1982 war and the 2006 military operations in Lebanon meanwhile were undertaken by Israel to quell the non-state threat from the PLO and the Hezbollah respectively. While the 1982 war succeeded in driving out the PLO from Beirut to Tunis, a more virulent anti-Israel organisation, the Hezbollah, took shape as a result of the continued Israeli presence in the country. By the time Israel withdrew in 2000, the Hezbollah grew from strength to strength, carrying out spectacular acts of terrorism, like the attack on the US military barracks in Beirut in 1983 and the bombing of the Israeli Embassy and Israeli cultural centre in Buenos Aires in 1992 and 1994 respectively.

*WMD Threats*

Great power politics and its acute sense of existential security have determined Israeli security choices vis-à-vis 'weapons of last resort'. In the aftermath of the 1956 Suez War, for instance, Israel was faced with an US arms embargo, while the Arabs began a rapid military build-up aided by the Soviet Union. Israeli leaders from David Ben-Gurion onwards have viewed nuclear weapons as the ultimate guarantor of their security. Analysts note that Ben-Gurion's nightmare was a second holocaust, this time at the hands of the Arabs.[20]

Some analysts believe that the decision to manufacture nuclear weapons was taken after the Six-Day War of June 1967. At the end of the war, Israeli forces captured stocks of Egyptian chemical weapons in Sinai. Israel's comprehensive victory in the war meanwhile failed to bring the Arab states to the negotiating table and to recognise its sovereignty. Analysts note that Israel's nuclear programme 'embodies its preference to maximize power and freedom of action' in an unstable neigbourhood.[21]

Israel is believed to have conducted a 'cold test' on 2 November 1966 and is therefore stated to have attained nuclear capability on that date, even though the test did not involve nuclear material, nor did Israel have the delivery system.[22] As for the size of the arsenal, Mordechai Vanunu, the technician who worked at the Dimona reprocessing plant, in his 1987 story to the London-based *Sunday Times* revealed that it produces 40 kilograms of weapons- grade plutonium in a year, sufficient to produce 100–150 weapons in the 1980s.[23]

The fundamental Israeli position on the question of nuclear weapons was first put forth by Shimon Peres, in a meeting with President John F. Kennedy on 2 April 1963, when he stated that Israel had no nuclear weapons and it would not be the first country to introduce nuclear weapons into the region.[24] Avner Cohen, in his important history of the Israeli nuclear weapons programme, notes that Prime Minister Levi Eshkol reiterated the Peres formulation in the Knesset on 18 May 1966. This was in the context of concerns expressed by Egyptian President Gamal Nasser on Israeli nuclear intentions.[25]

Cohen, therefore, argues that while Peres might have used the formulation for the first time, it was Eshkol who first publicly articulated it as Israel's declaratory policy. Prime Minister Yitzhak Rabin added in 1975 that Israel could not afford to be the second either to introduce nuclear weapons into the region.[26] Following the attack on the Osiraq reactor in Iraq in June 1981, Prime Minister Begin declared that Israel would prevent any attempt by its enemies to acquire nuclear weapons.[27]

The term that perhaps best describes the Israeli nuclear effort is 'opaque' nuclear proliferation. For Avner Cohen and Benjamin Frankel, Israel is the 'ideal type' of 'opaque' proliferation, in contrast to the US, which they characterise as the 'ideal type' of 'visible' proliferation. Israel has not conducted any nuclear tests, has consistently been insisting that it will not be the first to introduce nuclear weapons into the region, has not explicitly threatened the Arab states with nuclear weapons and does not have a military doctrine, like the US or any other overt nuclear power, that incorporates nuclear weapons into its overall security framework.[28]

An important factor that has fostered Israeli opacity was the need to prevent provoking the Arab states from getting nuclear weapons themselves. As long as the Arab states were not certain of the Israeli nuclear posture, it was believed they would not take any steps to further aggravate the situation. The Arab states, however, have embarked on their own WMD programmes, some of them not necessarily to offset Israeli nuclear capabilities. Iraq embarked on its nuclear programme in the late 1970s, for instance, in response to Iranian efforts in this arena, while Iran embarked on its missile programmes in the aftermath of the drubbing it received in the closing stages of the Iran–Iraq War from Iraqi ballistic missiles.

Since 2002 meanwhile, when the Iranian nuclear programme became a significant issue of international concern, Israel has been at the forefront of efforts to quell the challenge. Jerusalem contends that the Iranian nuclear ambitions, combined with its growing ballistic missile capabilities, pose an existential threat (see the sections on the Iranian nuclear challenge). Iranian economic and military support to NSAGs like the Hezbollah and the Palestinian Islamic Jihad, among others, have further fuelled Israeli angst against their state patron. The attack on the Syrian Al-Kibar under-construction nuclear facility in 2007—supposedly being built with North Korean assistance—is another proof of Israel's policy of not allowing any other country in its neighbourhood to have the wherewithal to attain a nuclear weapons capability.

## Case Studies

*2006 Israel–Hezbollah War*
The Lebanon War saw 34 days of fighting between Israel and Hezbollah which was triggered when the terror organisation kidnapped two Israeli soldiers after carrying out an attack on IDF Humvees on 12 July 2006. This attack on the Israel–Lebanon border happened exactly 19 days after the abduction of IDF soldier, Gilad Shalit, on the Gaza border by a Palestinian group. These abductions triggered a major domestic political crisis and pressures for swift retaliatory actions for the Ehud Olmert government, which decided to strike out hard against the Hezbollah with indeterminate war aims.

As soon as hostilities broke out, the Israeli Air Force destroyed large numbers of Hezbollah's stockpile of Fajr rockets (third generation Katyusha's), an artillery rocket with a range of 40 kilometre (km), within minutes in a precision bombing campaign. However, Hezbollah was able to fire over 4,000 Katyushas still left in its inventory throughout the time period of the war, completely paralysing the economic life in northern Israel. The IDF fighter jets also destroyed the Dahiya/Dahieh neighbourhood, Hezbollah's headquarters in the suburbs of Beirut. These actions by the IDF—which was headed by an Israeli Air Force General, Lieutenant General (Lt Gen) Dan Halutz, for only the second time in its history—however did little to destroy the organisation's fighting capabilities or its will to fight.

Analysts note that the military strength of the Hezbollah comprised of 10,000 fighters, over 13,000 107mm and 122mm short-range rockets (with a range of 20 km) and 1,000 Iranian and Syrian medium-range rockets (with ranges of over 100 km).[29] The Hezbollah carried out spectacular acts, like destroying an Israeli helicopter; destroyed scores of tanks and armoured personnel carriers by using advanced anti-tank missiles; and also damaged the warship, *INS Hanit*, which was anchored at Haifa by a missile strike.

Apart from the Shia organisation's military strengths and its fighting spirit, analysts note that Israel committed mistakes in not adequately preparing for a 'hybrid' conflict with Hezbollah. The IDF, for instance, privileged stand-off firepower, with the overwhelming intention being the reduction/prevention of casualties. The IDF was well-versed in counter-insurgency operations in the West Bank and the Gaza Strip, highly centralised actions carried out by infantry units and special forces, against Palestinian terror groups. Such tactics, however, were found to be woefully ineffective against an opponent like Hezbollah, armed with sophisticated weapons.

Decisions like the cut in budgets for training of reserve forces also impacted negatively on the IDF's performance. The IDF did not pay sufficient attention to active defence measures against short-range missile threats, which led to over 1 million Israelis living in bomb shelters during the duration of the month-long war, severely impacting the Israeli economy. Further, analysts like Efraim Inbar affirm that the political leadership wrongly assumed that Israeli public was tired of protracted conflict and hence was 'casualty-phobic'. Inbar instead argues that the leadership failed to appreciate that the public was 'defeat-phobic' and would have carried out everything necessary in order to achieve even expanded Israeli war aims, which in his view should have included severely hurting Hezbollah's state patron, Syria.[30]

The Hezbollah's robust media propaganda (with its own television stations like Al Manar) meanwhile was very effective. The organisation highlighted civilian casualties as a result of IDF air raids to domestic as well as international public opinion. Williamson Murray notes that while hybrid conflicts are the most difficult to win decisively, Israel 'suffered a political defeat of considerable proportions' during the 2006 Lebanon War.[31] The Winograd Commission constituted to do a post-mortem of the conflict concluded that the IDF 'failed...to provide an effective military response to the challenges posed to it by the war in Lebanon...'. The Commission found 'serious failings and shortcomings' in the decision-making process, in preparedness and training, in strategic thinking and planning, in efforts to defend civilian population, among others.[32]

After the war, General Halutz resigned in January 2007 after a critical report on his performance during the war. The IDF adopted a 'back-to-basics' approach, the training focus was shifted from LIC to HIC, the production of the Merkava main battle tank (MBT) was resumed and it learnt the lesson that 'stand-off precision fire was necessary but not a sufficient tactic' when the enemy resides among the population.[33] The IDF further placed greater emphasis on joint training and combat operations of different branches, strengthened its land manoeuvre capability and initiated training programmes for senior commanders, among other measures.[34]

## Gaza Conflicts Post-2005 Disengagement

Even prior to its disengagement from Gaza, Israel launched significant military interventions in response to rocket attacks. In September–October 2004, for instance, Israel launched a military offensive targeting Palestinian refugee camps in response to a rocket attack that killed two children in Sderot, bordering Gaza. Over 100 Palestinians were killed in this operation. After 2005, however, the cycle of violence increased dramatically. Over 400 Palestinians were killed in Israeli military operations ('Operation Summer Rains') in response to the kidnapping of Corporal Gilad Shalit in June 2006.

Major military interventions by Israel in the Gaza Strip since then have included 'Operation Cast Lead' (27 December 2008–18 January 2009), 'Operation Pillar of Defence' (14–21 November 2012) and 'Operation Protective Edge' (8 July—26 August 2014). The primary objective driving Israel's interventions was to stop the rocket attacks that progressively compromised its sense of security. The IDF contended that over 11,000 rockets were fired from the Gaza Strip between 2005 and July 2014, the beginning of Protective Edge.

Such periodic military actions (termed as 'mowing the grass' tactic) did recreate temporary deterrence from such provocations, though, despite suffering severe harm from the massive Israeli use of firepower, Hamas as well as Palestinian terror groups continued to replenish their capabilities. Israeli analysts however argue that the use of force was not designed to achieve 'impossible political goals' (like achieving mutually acceptable agreements with the Palestinian national movement which, in their view, is not ready/willing to make the necessary compromises for peace) but to degrade enemy military capabilities whose sole purpose was to wreck harm on Israel. Israel, therefore, had no choice but to engage in a 'war of attrition' with its enemies.[35]

The hugely disproportionate loss of Palestinian lives (over 4,000) as against about 100 Israelis cumulatively in these conflicts led to much international scrutiny about Israeli tactics. Israel, on its part, affirmed that the IDF took unprecedented steps to reduce civilian casualties and charged the Hamas with knowingly putting civilians in danger.[36] The IDF further insisted that the claim it used disproportionate force, as reflected in the huge differences in the numbers of fatalities on either side, reflects a:

> flawed understanding of the principle of proportionality under the Law of Armed Conflict, which requires a party planning an individual attack on a specific target to assess whether the expected incidental harm...would be excessive in relation to the concrete and direct military advantage anticipated.[37]

The IDF affirmed that in a situation of armed conflict, the amount of force used corresponds to that required to repel an armed attack.

Having learnt the importance of ground operations from the setbacks in the war with Hezbollah in 2006, the IDF in 'Operation Cast Lead' (2008–2009) used heavy forces comprised of two infantry, one armoured and one airborne infantry brigades to reduce operational risks and minimise friendly casualties.[38] In 'Protective Edge', the amount of firepower increased dramatically. Reports noted that 11 battalions of heavy artillery were used on a single day, 20 July 2014, to flatten the eastern Gaza neighbourhood of Shujaiya, which was teeming with Palestinian snipers and guerillas. This was in the aftermath of the death of 14 IDF soldiers (56 more were injured) as a result of an ambush by Hamas fighters—termed by critics as the 'greatest loss' suffered by the IDF in a short period of time.[39]

The IDF listed Hamas's rocket inventory to include the M-302 with a range of 160 km, the M-75 with a range of 75 km, the Grad rocket with a range below 50 km and the Qassam projectiles with a range below 20 km. In order to overcome Palestinian tactics which relied on such an inventory, active defence measures, ranging from missile defence to armour defence, gained prominence in the aftermath of the 2006 war. The IDF made good use of the Iron Dome anti-missile defence (AMD) system against short-range rockets fired from the Gaza Strip and the Trophy system to protect its MBTs/armoured personnel carriers (APCs).

During 'Protective Edge', for instance, even as nearly 4,600 rockets were fired into Israel, 800 projectiles deemed as a threat to Israeli population centres were intercepted by the Iron Dome system, with more than 90 per cent success rate. Israel, however, noted that over 3,800 of the rockets fired did land into Israel during this conflict, killing seven civilians and wounding 126.[40] The economic impact of the July–August 2014 intervention was pegged at nearly $1–2 billion.[41]

Apart from periodic heavy military interventions, economic warfare has been an important element of Israel's efforts to pressurise the Hamas and other groups from not carrying out disruptive attacks against it. One of the first strikes during 'Operation Summer Rains' in June 2006, in the aftermath of the Shalit kidnapping, for instance, was against the main power plant supplying electricity to Gaza. Israel also imposed intensified movement restrictions on Gaza from June 2007, in the form of land, sea and air blockade. These restrictions followed the violent takeover of the territory by Hamas from the Fatah-led Palestinian Authority (PA).

Exports from the Gaza Strip were completely banned in 2007, with limited relaxation allowed to permit the export of certain commodities like cut flowers beginning in 2010. Fishing restrictions continue to be in place, with Palestinians allowed to fish only up to 9 nautical miles (NM) as against the 20 NM allowed under the Oslo Accords.[42] Israel has also under taken punitive measures against international efforts to break the blockade, as evidenced in the Mavi Marmara incident in May 2010, which severely damaged Israel's relations with Turkey. Ten activists lost their lives in the raid. Israel subsequently apologized for the raid in March 2013 and reports noted that it agreed to pay compensation to those who had lost their lives. The United Nations (UN) Secretary-General, Antonio Guterres, visited the Gaza Strip in end August 2017 and urged for an end to the decade-long blockade.

Meanwhile, Israel on its part insists that the aim of the blockade is to prevent the supply of materials/equipment that could be used to make rockets/bombs. Further, Israeli analysts note that there are no restrictions relating to humanitarian aid/support, while water and electricity are also supplied. Electricity supply, however, is a subject of much contention. After Gaza's sole power plant stopped functioning in April 2017, for instance,

Israel is currently the sole supplier of electricity to an area housing 2 million people, with the quantum of its supply catering to the enclave's power needs for about 4 hours daily. Despite such limited supply, the PA in July 2017 urged Israel to further reduce power supply in order to put more pressure on Hamas, triggering a war of words between Hamas, PA and Israel.[43] Israel has also carried out the threat of withholding tax revenues it collects on behalf of the PA in response to such Palestinian actions like approaching the International Criminal Court to pursue charges of war crimes against it or for non-payment of electricity and water bills.

## Iranian Nuclear Challenge

Israel contends that Iran's nuclear ambitions combined with its hateful rhetoric (Israel being the 'Little Satan' in Tehran's terminology, combined with oft-repeated threats to 'wipe off' Israel from the map from President Mahoud Ahmadinejad, among others) constitute an existential threat. Ever since August 2002, when Iran's nuclear contentions occupied world attention, Israel has been very vocal in drumming up international support for stringent action to forestall the possibility of Tehran achieving its nuclear ambitions.

Prime Minister Benjamin Netanyahu, in his famous speech at the UN General Assembly in September 2012, in the context of quarterly reports by the International Atomic Energy Agency (IAEA) that Iran was rapidly expanding its stockpile of enriched uranium, charged that the diplomatic efforts of the international community for over a decade had failed. The Israeli prime minister equated the possibility of a nuclear-armed Iran as posing the same danger as a nuclear-armed Al Qaeda as 'both are fired by the same hatred… driven by the same lust for violence'.[44] In the same speech, Netanyahu told his audience that Iran's enrichment facilities 'are the only nuclear installations that we can definitely see and credibly target'.[45]

A few months earlier, in July 2012, Israel's Defence Minister, Ehud Barak, called for a 'swift and definite stop' to the Iranian nuclear project.[46] Barak, in August 2015, revealed that Israel had considered attacking Iran at least three times between 2009 and 2012, but did not do so either because the IDF's assessment was that it did not have the requisite operational capabilities to carry out successful strikes or because senior members of the Netanyahu cabinet disagreed over such a course of action.[47] Israeli diplomatic pressure for the exercise of military option however did not fall on receptive ears of the Barack Obama administration, which continued to privilege its 'dual-track' policy of 'applying pressure in pursuit of a constructive engagement, and a mutually acceptable solution'.[48]

Apart from such robust diplomatic pressure, Israel did in fact carry out destabilising activities designed to hurt the Iranian nuclear programme, which of course it does not confirm or deny. The Stuxnet virus is a prominent example, which has been termed as the 'world's first digital weapon'.[49] Beginning from June 2009, the virus affected P-1 centrifuges at the Natanz uranium enrichment plant, leading to their malfunctioning. The virus is believed to be the result of joint collaboration between the National Security Agency (NSA) of the US and Unit 8200, the crack cyber unit of the IDF.[50]

The death of Iranian nuclear scientists in unexplained car bomb explosions is another tactic that Israel is alleged to have been involved in order to set back the country's nuclear programme. At least five nuclear scientists were killed till 2015. Israeli Defence Minister Moshe Yaalon, in an interview with a German newspaper, when asked whether Israel was responsible for carrying out such attacks, insisted that 'one way or another, Iran's military nuclear program must be stopped. We will act in any way and are not willing to tolerate a nuclear-armed Iran ....'[51]

## In Closing

This chapter shows that Israel has faced a unique security situation, with the threat spectrum ranging from state-centric threats (with opponents using regular and irregular means at their disposal) to non-state threats (with or without external state sponsors) as well as WMD threats. Israel was able to tackle such threats effectively whenever it adopted a use of force doctrine/posture commensurate with the strategic reality. This held true for the strategic defensive posture, with the offensive armoured manoeuvre being the main tactic to tackle the invading Arab armies.

Offensive armoured divisions are, of course, of not much use against the 'small war' tactics of the Palestinian groups. Various tactics like 'mowing the grass'—the closest approximation to an Israeli 'lighthouse' theory on fighting hybrid opponents[52]—among others, including economic warfare, were employed to counter the challenge from the 'occupied territories'. The training or preparations designed primarily to deal with such threats as those emanating from the Palestinians however proved inadequate while dealing with hybrid opponents like the Hezbollah.

As seen during the 2006 war with Hezbollah, while Israel struggled to mount effective combined forces operations integrating armoured and artillery forces with air strikes, its hybrid opponent inflicted heavy damage using sophisticated anti-tank missiles like RPG-29 and Konkurs and Kornet, along with anti-ship missiles like C-802, which hit *INS Hanit*. While the lack of clarity as regards war aims further muddied the situation for Israel right from the initial outbreak of hostilities, Hezbollah mounted an effective information operations campaign highlighting civilian casualties due to the IDF air strikes.

Learning from its experience in the 2006 war with the Hezbollah, the IDF used massive conventional firepower during the Gaza conflicts post-disengagement. This was done to not only reduce casualties among its ranks but to convey an unambiguous message of its unvarnished intention to severely degrade enemy capabilities, as well as its infrastructure and population centres, in the hope of ensuring deterrence. These conflicts have, however, occurred at dangerously regular intervals and have only ensured limited deterrence against Palestinian armed groups. Use of active defence measures like the Iron Dome AMD system meanwhile have brought a significant sense of security and respite from debilitating rocket attacks.

With the Palestinian terror groups continuing to work on their options to hurt Israel, including by the building of underground tunnels to infiltrate into Israel and carry out terror attacks, Israel is investing in technological solutions like underground fences equipped with motion detection sensors to prevent such instances of infiltration. Lone wolf attacks are also a reality that Israeli security forces have to grapple with. Given the above, the next round of renewed and expanded violence seems eerily around the corner. Israeli analysts, on their part, note that the country has no choice but to be prepared for a long and hard 'war of attrition' with the Palestinians, in the context of the still significant political divides.

While the immediacy of the threat from the Iranian nuclear programme has reduced after the Joint Comprehensive Plan of Action (JCPOA), Israel continues to be wary of Tehran's nuclear intentions. Israeli political and military leaders however insist that if required, they would not be found wanting with regard to operational capabilities to significantly set back the Iranian nuclear programme anytime in the future if the situation so demands, contingent upon Iranian nuclear behaviour. President Donald Trump not recertifying the JCPOA in mid-October 2017 meanwhile has added to uncertainties regarding the future implementation of the deal.

Iran's continuing testing of rockets (with 'death to Israel' scribbled across them) and its undiminished support to the Hamas, Hezbollah and Palestinian Islamic Jihad riles Israel. Analysts believe the Hezbollah has replenished its inventory with more sophisticated rockets and is raring for a fight, as and when the opportunity arises again. While Israel would seem better prepared the next time around with technological solutions pertaining to active defence systems like the Iron Dome and its larger 'cousin', the David's Sling, deployed, it will again face the same dilemmas with respect to the nature of its war aims in a future conflict with the Hezbollah.

Israel, therefore, as has been shown in the three case studies, has been the target of hybrid tactics (both of the kinetic and non-kinetic variety) and has also used similar tactics to offset its opponents, with varying degrees of success. The Israeli experience with the Hezbollah and Hamas highlights the relevance of effective combined arms conventional kinetic capabilities to tackle the significant conventional wherewithal that hybrid groups like the Hezbollah and Hamas bring into the hostile equation. The 2006 Hezbollah War is a cautionary tale of the continuing relevance of effective training and preparations that need to be made as regards the conventional capabilities of a nation-state, against enemy capabilities that range from the kinetic to the non-kinetic spectrum. The Lebanon War is also illustrative of the fact that while Israel has been innovative throughout its history in executing successfully tactics specific to a particular security situation—across the three dimensions of state-centric, non-state as well as WMD threats—it was forced to re-evaluate its use of force philosophy and practice in the face of setbacks as in Lebanon.

As for the applicability of the Israeli scenario to other security contexts like that being faced by India, differences in security contexts and the nature of the threat perceptions, among others, preclude the automatic transplantation of any single kinetic/non-kinetic response. Tactics like the targeted assassination of terrorist masterminds, for instance, have not always been successful for the Israeli security agencies. The botched attempt to kill Khaled Meshaal on the streets of Amman in 1997 is an example, even as the 1976 Entebbe raid straddles the other end of the spectrum as the epitome of successful special forces operations. Israel's effective use of technology in the form of active defence systems and perimeter security systems is worthwhile to be emulated. What is important to bear in mind is the extent to which any tactic comports with the larger strategic objective of a particular security situation.

## NOTES

1. T.X. Hammes, 'The Art of Petraeus', *The National Interest*, No. 98, November–December 2008, p. 56.
2. Frank G. Hoffman, 'Hybrid Threats: Re-conceptualizing the Evolving Character of Modern Conflict', *Strategic Forum*, No. 240, April 2009, p. 5, available at https:/

/www.files.ethz.ch/isn/98862/SF240.pdf, accessed on 30 August 2017.
3. See Frank G. Hoffman, 'Hybrid Warfare and Challenges', *Joint Forces Quarterly*, No. 52, First Quarter 2009, p. 36. For Hoffman, in 'compound wars', irregular forces 'operated as a distraction or economy of force measure in a separate theatre or adjacent operating area...'. Hybrid warfare, however, involves the operational fusion of conventional and irregular capabilities not just at the strategic but at the tactical level as well. See also Peter R. Mansoor, 'Introduction: Hybrid Warfare in History', in Williamson Murray and Peter R. Mansoor (eds), *Hybrid Warfare: Fighting Complex Opponents from the Ancient World to the Present*, New York: Cambridge University Press, 2012, pp. 1–17.
4. David E. Johnson, 'Military Capabilities for Hybrid War: Insights from the Israel Defense Forces in Lebanon and Gaza', RAND Occasional Paper, 2010, p. 2, available at https://www.rand.org/content/dam/rand/pubs/occasional_papers/2010/RAND_OP285.pdf, accessed on 13 February 2017.
5. See Chapter 2, 'Contextualising Hybrid Wars', by Vikrant Deshpande and Shibani Mehta.
6. Efraim Inbar, *Israel's National Security: Issues and Challenges since the Yom Kippur War*, London: Routledge, 2008, p. 56. See also J.S. Levy and J.R. Gochal, 'Democracy and Preventive War: Israel and the 1956 Sinai Campaign', *Security Studies*, Vol. 11, No. 2, September 2010, pp. 1–49.
7. Fuad Jabber, *Israel and Nuclear Weapons: Present Options and Future Strategies*, London: Chatto and Windus, 1971, p. 106.
8. Israel Tal, *National Security: The Israeli Experience*, translated by Martin Kett, London: Praeger, 2000, p. 37.
9. Ibid., p. 43.
10. David Rodman, 'Regime Targeting: A Strategy for Israel', in Efraim Karsh (ed.), *Between War and Peace: Dilemmas of Israeli Security*, London: Frank Cass, 1996, p. 153.
11. Inbar, *Israel's National Security*, n. 6, p. 22.
12. Shimon Naveh, 'The Cult of the Offensive: Pre-emption and Future Challenges for Israeli Operational Thought', in Karsh (ed.), *Between War and Peace*, n. 10, p. 170.
13. Rodman, 'Regime Targeting', n. 10, p. 156.
14. See https://www.shabak.gov.il/english/publications/Pages/terror-organizations/terror-organizations.aspx, accessed on 20 August 2017.
15. Edward E. Azar, Paul Jureidini and Ronald McLaurin, 'Protracted Social Conflict: Theory and Practice in the Middle East', *Journal of Palestine Studies*, Vol. 8, No. 1, Autumn 1978, pp. 41–60.
16. Inbar, *Israel's National Security*, n. 6, p. 39.
17. Ministry of Foreign Affairs (MFA), 'In Memory of the Victims of Palestinian Violence and Terrorism in Israel', available at http://mfa.gov.il/MFA/ForeignPolicy/Terrorism/Victims/Pages/In%20Memory%20of%20the%20Victims%20of%20Palestinian%20Violence%20a.aspx, accessed on 20 August 2017.
18. Inbar, *Israel's National Security*, n. 6, p. 196.
19. Boaz Ganor, 'Israel, Hamas and Fatah', in Robert J. Art and Louise Richardson (eds), *Democracy and Counter-Terrorism*, Washington, DC: USIP Press, 2007, pp. 294–297.
20. Seymour M. Hersh, *The Samson Option: Israeli Nuclear Arsenal and American Foreign*

21. Inbar, *Israel's National Security*, n. 6, p. 88.
22. Michael Karpin, *The Bomb in the Basement: How Israel went Nuclear and What that Means for the World*, New York: Simon and Schuster, 2006, p. 268.
23. Frank Barnaby, *The Invisible Bomb: The Nuclear Arms Race in the Middle East*, London: I.B. Tauris and Co. Ltd, 1989, p. 25; Hersh, *The Samson Option*, n. 20, p. 197.
24. Hersh, *The Samson Option*, n. 20, p. 119.
25. Avner Cohen, *Israel and the Bomb*, New York: Columbia University Press, 1998, p. 233.
26. Frank Barnaby, 'Capping Israel's Nuclear Volcano', in Karsh (ed.), *Between War and Peace*, n. 10, p. 105.
27. Rodney W. Jones, Mark G. McDonough, Toby F. Dalton, and Gregory D. Koblentz, *Tracking Nuclear Proliferation*, Washington, DC: Carnegie Endowment for International Peace, 1998, p. 206.
28. Avner Cohen and Benjamin Frankel, 'Opaque Nuclear Proliferation', in Benjamin Frankel (ed.), *Opaque Nuclear Proliferation: Methodological and Policy Implications*, London: Frank Cass, 1991, pp. 14–44.
29. Eleazar S. Berman, 'Meeting the Hybrid Threat: IDF's Innovation against Enemies 2000–2009', 16 April 2010, available at https://repository.library.georgetown.edu/bitstream/handle/10822/553449/bermanEleazar.pdf, accessed on 20 August 2017.
30. Inbar, *Israel's National Security*, n. 6, pp. 228–231.
31. Williamson Murray, 'Conclusion: What the Past Suggests', in Murray and Mansoor (eds), *Hybrid Warfare*, n. 3, p. 290.
32. See 'Winograd Commission: Interim Report', 30 April 2007 and 'Winograd Commission: Final Report', 30 January 2008, available at http://www.jewishvirtuallibrary.org, accessed on 25 August 2017.
33. Johnson, 'Military Capabilities for Hybrid War', n. 4, pp. 4–5.
34. See 'Winograd Commission: IDF Response to the Final Report', 30 January 2008, available at http://www.jewishvirtuallibrary.org/idf-response-to-the-winograd-commission-final-report-january-2008, accessed on 25 August 2017.
35. Efraim Inbar and Eitan Shamir, '"Mowing the Grass": Israel's Strategy for Protracted Intractable Conflict', *Journal of Strategic Studies*, Vol. 37, No. 1, 2014, pp. 67–90.
36. See MFA, '2014 Gaza Conflict: Factual and Legal Aspects', May 2015, available at http://mfa.gov.il/ProtectiveEdge/Documents/2014GazaConflictFullReport.pdf, accessed 25 August 2017. See also IDF, 'Operation Protective Edge', *IDF Blog*, available at https://www.idfblog.com/operationgaza2014/#Terror, accessed on 25 August 2017.
37. MFA, '2014 Gaza Conflict', n. 36, p. 55, footnote 163.
38. Johnson, 'Military Capabilities for Hybrid War', n. 4, pp. 6–7.
39. Max Blumenthal, *The 51 Day War: Ruin and Resistance in Gaza*, New Delhi: Left Word Books, 2015, p. 42.
40. MFA, 'Operation Protective Edge: Israel under Fire, IDF Responds', 26 August 2014, available at www.mfa.gov.il, accessed on 25 August 2017.
41. MFA, '2014 Gaza Conflict', n. 36, pp. 132–135.
42. See United Nations Office for the Coordination of Humanitarian Affairs, 'Gaza

Blockade', available at https://www.ochaopt.org/theme/gaza-blockade, accessed on 25 August 2017.

43. Adam Ragson, 'Palestinian Energy Authority: Israel begins to Reduce Gaza Electricity', *The Jerusalem Post*, 19 June 2017, available at http://www.jpost.com/Arab-Israeli-Conflict/Israel-reduces-electricity-to-Gaza-strips-Energy-Authority-says-497298, accessed on 27 August 2017.

44. 'PM Netanyahu Addresses UN General Assembly', 27 September 2012, available at http://embassies.gov.il/MFA/PressRoom/Pages/PM-Netanyahu-addresses-UN-General-Assembly.aspx, accessed on 29 August 2017.

45. Ibid.

46. "Iranian Nuke Threat Outweighs Cost to Block Progress, Israel Says", 26 July, 2012, available at http://www.nti.rsvp1.com/gsn/article/iranian-nuke-threat-outweighs-cost-blocking-progress-israel/?mgh=http%3A%2F%2Fwww.nti.org&mgf=1 (accessed on August 2, 2012.

47. Barak Ravid, 'Barak: Steinitz, Ya'alon Thwarted Iran Strike in 2011', *Haaretz*, 23 August 2015, available at http://www.haaretz.com/israel-news/1.672335, accessed on 29 August 2017.

48. US Department of State, 'Joint Statement on Iran Sanctions', 23 June 2011, available at http://www.state.gov/r/pa/prs/ps/2011/06/166814.htm, accessed on 24 January 2012.

49. Kim Zetter, 'An Unprecedented Look at Stuxnet, the World's First Digital Weapon', 3 November 2014, available at https://www.wired.com/2014/11/countdown-to-zero-day-stuxnet/, accessed on 30 August 2017.

50. David E. Sanger, 'Obama Order Sped up Wave of Cyberattacks against Iran', *The New York Times*, 1 June 2012, available at http://www.nytimes.com/2012/06/01/world/middleeast/obama-ordered-wave-of-cyberattacks-against-iran.html?pagewanted=1&_r=1&hp, accessed on 30 August 2017.

51. 'Israel behind Assassinations of Iran Nuclear Scientists, Ya'alon Hints', *The Jerusalem Post*, 7 August 2015, available at http://www.jpost.com/Middle-East/Iran/Israel-behind-assassinations-of-Iran-nuclear-scientists-Yaalon-hints-411473, accessed on 30 August 2017.

52. Vikrant Deshpande suggested that Israel's 'mowing the grass' tactic best corresponds to an Israeli 'lighthouse' theory on hybrid warfare.

# 8

# Expanding the Turbulent Maritime Periphery
## Gray Zone Conflicts with Chinese Characteristics

*Abhay K. Singh*

During the Raisina Dialogue 2018, Admiral Harris termed China as a 'disruptive transitional force in the Indo-Pacific'.[1] China's aggressive maritime irredentism is one of its key disruptive characteristics. Even though China dominates the Pacific geographically, the maritime milieu in the Pacific is grim for China. Its Pacific coast, till the first island chain, consists of disputed territories, potentially a serious obstacle in China's unhindered access to the larger ocean.[2] China is party to multiple, decades-old maritime territorial disputes, which have periodically led to incidents and episodes of increased tension. In the past decades, China has used a spectrum of assertive strategic tools to progressively shift the status quo in its favour.[3] China's claims and presence in the East and South China Seas has been growing and is seemingly permanent.

Chinese assertive activities in the region, especially in the East China Sea and South China Sea, display a strategic approach which bears clear attributes of 'gray zone' conflict strategy as described in Chapter 2. As per Hal Brands, the aim of 'gray zone' approaches:

> is to reap gains, whether territorial or otherwise, that are normally associated with victory in war. Yet gray zone approaches are meant to achieve those gains without escalating to overt warfare, without crossing established red-lines, and thus without exposing the practitioner to the penalties and risks that such escalation might bring.[4]

China has thus far advanced its claim through a variety of measures, political, diplomatic and military, including coercion of potential opponents through limited escalation and delayed resolution of issues which it could not settle in its favour as yet. While it has refrained from launching large-scale aggression or an all-out military operation, China has pursued a strategy of small, incremental but persistent enhancement of its jurisdiction claims in the East China Sea and the South China Sea. In view of Michael Mazzar, Chinese conceptions of strategy are inherently attuned to gray zone approaches.[5] China's preference for a long-term indirect approach to unobtrusively manoeuvre the strategic configuration of the region in its favour is grounded in classic Chinese thought. According to Frank Hoffman, 'China's diplomatic assertions, information announcements, and deliberate use of fishery/maritime security forces to assert sovereignty in and around contested shoals and islands in the Pacific constitute a good case study in deliberately deniable acts of aggression.'[6]

The chapter details the various elements of China's strategic approach in the maritime gray zone conflicts of the East and South China Seas. It first explores China's classic and contemporary strategic literature and highlights relevant strategic concepts. Then, China's coercive gray zone approaches in the East and South China Seas have been examined. The last part assesses effectiveness of the Chinese gray zone approaches and its future contours.

## Exploring China's Gray Zone Strategic Concepts

Stratagem and deception define the key essence of the gray zone strategic concept. With the world's first comprehensive military classic, *The Art of War*, and the largest number of ancient military writings, China considers itself the birthplace of stratagems. While the ancient Chinese military classics cover a wide spectrum of the issues related to warfare, strategy and stratagems remains their key focus area. Henry Kissinger has noted this Chinese preference for stratagems and indirect approach in conflicts and diplomacy as opposed to the Western strategic traditions. In his view, 'whereas Western tradition preferred the decisive clash of force emphasising feat of heroism, the Chinese ideal stressed subtlety, indirection and patient accumulation of relative advantage.'[7]

### A Wei Qi Approach to Gray Zone Conflict

The ancient Chinese game of 'Wei Qi', also popular in Japan as 'Go', provides a compelling insight into China's gray zone conflict strategy. As a game of strategy, Wei Qi or Go is part of Chinese strategic culture. It takes Chinese philosophical and military thinking as its foundation and puts Chinese strategic thinking and military operational art into play.[8] The game is played on a board with 19×19 matrix by two players who sequentially place black and white stones at intersections. The aim is to engage each other through moves of stones and capture territories through penetration. The black and white exemplify the concept of yin and yang and penetrating moves of players as the flow of water. The player acquiring more territories is considered the winner. The players' gambit during the game aim for territorial conquest through their placement of stones imitate warfighting strategies of invasion, engagement, confrontation and encirclement. As per David Lai, 'Sun Tzu's thoughts and the essential features of the Chinese way of war are all played out in the game.'[9] As the game unfolds, it becomes a war with multiple campaigns and battlefronts. Seen from the perspectives of international politics, it is a competition between two nations over multiple areas of interest.

In many ways, Wei Qi embodies the key dictum of Sun Tzu, as the guiding principle of the game is Sun Tzu's truism about subjugating the enemy without fighting. The player's moves in the game of Wei Qi embodies Sun Tzu's precepts: first, by frustrating the enemy's strategy; then by derailing its allies; and finally, by attacking the enemy's military in order defeat the opponent.[10] It has been argued that the strategic precept of Wei Qi has an immense impact on the Chinese strategic behaviour in international conflicts, and makes the Chinese way of war different from those of other cultures.[11]

China's strategic thinking is, by design and history, much more comprehensive and diverse. The game of Wei Qi signifies key strategic approaches in gray zone conflicts. Since 1978, when China started the mission of national development, there has been renewed interest in the intellectual exploration of classical military literature and Sun Tzu. Utilisation of the distilled wisdom of its classic strategic thought in pursuing its national objective is also becoming increasingly evident.

*Gray Zone Stratagem in Chinese Military Classics Literature*
In Chinese history, elements of hybrid warfare have often been crucial components of conflicts with their neighbours. Ancient Chinese rulers essentially employed four methods to pacify their troubled boundary.[12] The first approach was of 'using barbarian to fight barbarians'. In this, by using barbarian mercenaries and strategic alliances, the neighbours would be kept divided. The contemporary analogy would be 'diplomatic warfare', that is, neutralising unfriendly states through public diplomacy and creating

fissures among alliances and partnership. The second approach was bribes and tributes to challengers in order to dissuade them from attacking China. The current approach of lucrative trade and aid deals is, in essence, inducement. The third approach was strong fortification, which China built to dissuade external offensive design. Recourse to military expedition was a final gambit when all other efforts did not yield the desired results.

The ancient Chinese values and warfighting principles remain relevant even today for Chinese policymakers and military elites. *The Seven Military Classics of Ancient China* is a wide-ranging and remarkably heterogeneous collection of strategic lessons accumulated during 'warring states' period and still remains a key reference material in the education of political and military elites in China.[13] Although Sun Tzu's *Art of War* is one of the most popular books outside China, the *Wu-Zi* and *Six Secret Teachings* have proven to be highly important sources for military wisdom over the centuries, with the latter continuing to be held in high esteem among contemporary People's Republic of China military professionals. The classic military literature provides tenets covering whole spectrum of military operations:

> from simple tactical principles through complex methods of organization and encompass extensive materials on command and control, campaigning, psychological operations and disinformation, manoeuvre, strategic power, intelligence, manipulating the enemy, deception, regulation and constraint, evaluating the enemy, mustering martial '*shi*', and the very nature of warfare itself.[14]

One of the key aspects of Chinese stratagem is *shi*, which is often translated as the 'propensity of things' which a general must aim to exploit to his own advantage and to maximum effect whatever conditions he encounters. The constant search for a strategic advantage or *shi* is also a goal of the Chinese strategic game of Wei Qi or Go. *Shi* is sought everywhere, whether it be with the use of forces or some other aspect of the strategic environment. The concept of *shi*, in essence, is metaphysical and difficult to accurately transcribe due to lack of an equivalent term in Western literature. Timothy Thomas has argued that '*Shi* is the goal of strategy's objective and subjective aspects: to create and attain an advantage over an opponent after evaluating a situation and influencing it.'[15] The United States (US) Department of Defense, in its annual report citing Chinese linguists, has explained it as 'the alignment of forces', the 'propensity of things' or the 'potential born of disposition' that only a skilled strategist can exploit to ensure victory over a superior force. It has been further argued that only a sophisticated assessment by an adversary can recognise the potential exploitation of *shi*.¹⁶ Timothy L Thomas considers that 'the posture of the army, strategic advantage, a strategic configuration of power, the alignment of forces and availing oneself of advantage to gain control are all used to define *shi*.'¹⁷ A significant component of *shi* is called *wuwei*, which means to get other nations to do work for you. *Shi* is also about taking and maintaining the initiative. As Sun Tzu puts it, 'those skilled at making the enemy move do so by creating a situation to which he must conform.'¹⁸ Michael Pillsbury considers two elements critical to the contemporary Chinese strategy: deceiving others into doing your bidding; and waiting for the point of maximum opportunity to strike.¹⁹

Inherent in the Chinese military strategic culture is the enduring believe in efficacy of secrecy and stratagem in obtaining a decisive victory against even a superior adversary. As Sun Tzu put it, 'All warfare is based on deception.'²⁰ The clear objective of the Chinese approach to warfare is to induce cognitive confusion in the adversary's mind through effective psychological operation. A quick and decisive operation, executed with precision, aims to confound the adversary both materially and psychologically.

*Strategy of Gray Zone Conflict: Contemporary Concepts*
Contemporary Chinese doctrinal writings appear to bear the imprint of its ancient philosophical legacy.²¹ Authors of the Chinese handbook, *The Science of Military Strategy*, argue that 'Strategic thought is always formed on the basis of certain historical and national cultural tradition, and formulation and performance of strategy by strategists are always controlled and driven by certain cultural ideology and historical cultural complex.'²² This section explores the strategy of gray zone conflicts in China's contemporary strategic concepts.

Insofar as gray zone strategy is concerned, the concept of 'unrestricted warfare', propounded by two Chinese Colonels, is in essence a manifesto of hybrid warfare on steroids.²³ The main arguments in the book are based on the premise that in unrestricted warfare, there are no rules and the boundary between battlespace and non-battlespace is progressively blurring. Unrestricted warfare argues for the overcoming of boundaries, restrictions and even taboos that separate the military from the non-military, the weapon from the non-weapon, and military personnel from non-military personnel. The 'non-military war operations' listed in the book include a variety of resources which have hitherto remained outside the military realm. The variety of warfare methods listed by the authors includes trade war, financial war, resource warfare, economic aid warfare, smuggling warfare, fabrication warfare, ecological warfare, network warfare, technological warfare, new terror war, use of 'hyper-strategic weapons' which in some ways replace nuclear devices, media warfare, psychological warfare, drug warfare, cultural warfare and international law warfare. Unrestricted, however, does not mean unlimited methods. Unrestricted warfare has been termed as 'combination warfare' as it brings into play new combinations of resource and methods and these include: supra-domain combinations (combining battlefields and choosing the main domain); supra-means combinations (combining all available means, military and non-military, to carry out operations); supra-national combinations (combining national, international and non-state organisations to a country's benefit); and supra-tier combinations (combining all levels of conflict into each campaign).

As a strategic theory, intellectual relevance of unrestricted warfare lies in the fact that it stands at the intersection of both Western and Eastern strategic culture. While the concept sources recent strategic experiences of the Western world, particularly means and methods employed in the First Gulf War, among others, 'in essence it is a post-modern version of Sun-Tzu's Art of War, updated to take into account not only the digitalization of the battlefield but the weaponization of law and the financialization of foreign policy.'[24]

The Chinese government has refused to acknowledge the book as representative of its thinking. Nonetheless, many experts consider it as a virtual blueprint for China's future war. Even if the book was to be considered as a combined opinion of two middle-ranking officers of People's Liberation Army (PLA), the concept has found wide appeal in China. The book is a bestseller in China and apparently remains in high demand, with reports of two pirated editions in circulation, in addition to multiple official editions. More importantly, President Jiang Zemin and the Defence Minister, Chi Haotian, are said to have read the book with great interest.[25]

Despite official disavowal of the unrestricted warfare precepts, the Chinese leadership has progressively internalised the value not only of military operations but also of non-military operations in order to win contemporary wars and conflicts, and furthermore to become a global superpower.[26] Based on the evaluation of the Second Gulf War by Central Military Commission (CMC), Jiang Zemin pointed out:

> the conduct of public opinion warfare, psychological warfare, and legal warfare by the use of modern mass media is an important

measure for warring countries which attempt to grasp the political initiative and military victory. We need to lead public opinion warfare, psychological warfare, and legal warfare into an important position in order to adapt to new circumstances.[27]

In December 2003, the Communist Party of China (CPC) Central Committee and the CMC approved the concept of 'three warfares' and promulgated the revised PLA 'Political Work Regulations', assigning the PLA to conduct 'a function for political work operations' through 'public opinion, psychological, and legal warfare'.[28] China's defence white paper in 2006 accordingly highlighted that PLA will:

> upgrade and develop the strategic concept of people's war, and work for close coordination between military struggle and political, economic, diplomatic, cultural and legal endeavours, uses strategies and tactics in a comprehensive way, and takes the initiative to prevent and defuse crises and deter conflicts and wars.[29]

The three warfares strategy, through the implementation of non-kinetic, non-violent, but still offensive operations, is best suited for Chinese peacetime strategy of influencing the cognitive processes of a country's leadership and population, or what Sun Tzu describes as 'subduing the enemy without fighting'.[30] A study commissioned by the US Department of Defense concludes that 'if the object of war is to acquire resources, influence and territory, and to project national will...*China's Three Warfares is war by other means.*'[31] The key components of three warfares are as follows:

- *Psychological warfare* seeks to influence and/or disrupt an opponent's decision-making capability, to create doubts, foment anti-leadership sentiments, to deceive opponents and to attempt to diminish the will to fight among opponents. It employs diplomatic pressure, rumour, false narratives and harassment to express displeasure, assert hegemony and convey threats.
- *Media warfare* (also known as *public opinion warfare*) is a 'constant, ongoing activity aimed at the long-term influence of perceptions and attitudes'.[32] It leverages all instruments that inform and influence public opinion, including films, television programmes, books, the Internet and the global media network (particularly *Xinhua* and *CCTV*) and is undertaken nationally by the PLA, locally by the People's Armed Police, and is directed against domestic populations in target countries.
- *Legal warfare* (or 'Lawfare') exploits the legal system to achieve political or commercial objectives.[33] Legal warfare uses the international and domestic law to claim the legal high ground or

assert Chinese interests. It can be employed to constrain operational freedom of the enemy and for shaping the operational space for own strategic advantage. Building international support and management of likely adverse political consequences of China's military actions are also intended purposes of law warfare.

The PLA's operational hierarchy of combat is divided into three levels: war, campaigns and battles. Each is informed by its own distinct level of operational guidance: strategy, campaign methods and tactics, respectively. The three warfares are primarily classed as a campaign method, but with the additional application at the strategic and tactical levels. Furthermore, the use of the three warfares reflects the PLA's underlying belief that war is not simply 'a military struggle, but also a comprehensive engagement proceeding in the political, economic, diplomatic and legal dimensions'.[34] The PLA officers' curricula include formal academic discourse on public opinion warfare, psychological warfare and legal warfare, even though the three warfares is a recent addition to the Chinese strategy.[35]

Reports indicate that the PLA Nanjing International Relation College examined the concept of 'disintegration warfare' between 2003 and 2009.[36] The college is part of the Intelligence Department of General Staff Headquarters and conducts training for military attachés and intelligence officers prior to their deputation in China's embassies around the world. According to this research, 'disintegrating enemy forces' is one of the three major principles prescribed by Mao Zedong as political work to be carried out by the Communist Party. Even today, it is considered an unused and potentially valuable strategy for the PLA. At its core lies Sun Tzu's dictum that 'to subdue the enemy without fighting is the supreme excellence'— which is to say, it is about winning without fighting. The academy's research has sought to keep this tradition of disintegration warfare alive, examining ways of 'winning without fighting' in a modern-day context by taking in the lessons learned from 'informatised' wars fought in recent years.[37] Disintegration warfare comprises activity in a variety of arenas: political, media, economics, psychology, information and strategy. It needs to be noted that warfare over media and economics are two kinds of disintegration warfare that begin during peacetime.

## China's Gray Zone Approach in the Maritime Disputes

An American diplomat famously said that the great powers do not go to war over rocks.[38] However, maritime disputes in East Asia and South China Sea are about controlling a few barren islands and low tide elevations.

Unarguably, these features are not barren rocks for the parties to the dispute; and these territorial disputes underline their sovereignty, resource and strategic significance for the countries concerned. These are long-standing complex, legal disputes, with overlapping claims and counter-claims for jurisdictions, with no peaceful settlement yet in sight. China, through astute use of gray zone tactics, has advanced her territorial claims in the maritime arena without triggering crises, escalations or conflicts—and thus without breaching the kinds of red lines that, when crossed, might have drawn in the US.[39]

## Gray Zone Dynamics in East China Sea

In many ways, East China Sea is an enclosed sea, surrounded on the west by the China mainland and in the east by Japan's southern island of Kyushu, the Ryukyu Island chain and Taiwan—a significant portion of the so-called first island chain. East China Sea is also the nexus of Sino-Japanese distrust and strategic competition. In recent years, incidents around the Senkaku Islands have become the main source of tension between the two countries. The Senkaku/Diaoyu Islands comprise five islets and three rocks in East China Sea about 200 kilometre (km) northeast of Taiwan, 400 km west of Okinawa and 300 km east of the Chinese mainland. The islands cover a small area of about 6 square km, but this belies their economic and strategic importance. They have a bearing on the dispute between Japan and China over their maritime border in the East China Sea and resultant rights over economic resources. Tokyo and Beijing have built their respective claims by cherry-picking aspects of historical record. In general, Japan relies on modern notions of international law, while China's case rests on concepts of historical title. China attaches great importance to the period of Chinese initial discovery of the islands, whereas Japan stresses its decades of unchallenged administration.

From the 1970s till the 1990s, Senkaku dispute largely remained dormant due to an unofficial understanding between China and Japan to shelve the dispute in favour of better economic and diplomatic relationship.[40] While Chinese fishing boats had been fishing in the waters around Senkaku since the 1970s, violation of the Japanese exclusive economic zone (EEZ) began in the 1990s. Also, the trend of challenging Japan's *de facto* control of the islands started as early as 2004 when seven Chinese activists landed on Uotsuri/Diaoyu Dao using fishing boats. Japan protested diplomatically and the Japanese police arrested the Chinese activists. However, they were not prosecuted and were released due to diplomatic pressure from China.[41]

In February 2007, a Chinese research ship was detected in the vicinity of the Senkaku Islands. Japan's diplomatic protest was disregarded by China with the assertion that the Diaoyu Islands were part of Chinese territory. The situation deteriorated in 2008 when the Japanese Coast Guard (JCG) spotted two ships of the China Marine Surveillance force in the Senkaku/Diaoyu Islands area. As per Chinese officials, the presence of Chinese government vessels was aimed at asserting its jurisdiction and for accumulating a record of effective control.[42] This Chinese approach is known as 'cabbage strategy', which entails surrounding a contested area with so many boats—fishermen, fishing administration ships, marine surveillance ships, navy warships—that 'the island is thus wrapped layer by layer like a cabbage'. China has displayed this strategy in enhancing its control both in East China Sea and South China Sea.[43]

The September 2010 arrest of a Chinese trawler captain near the islands raised the temperature still further. The trawler was operating very close to the islands and had attempted to ram two JCG vessels. The trawler captain was subsequently detained and charged under Japanese law. In addition to diplomatic protests, China, in retaliation, suspended exports of rare earth metals to Japan, cancelled a second round of talks about energy cooperation in East China Sea and postponed the visit of 1,000 Japanese youth to the Shanghai World Exhibition. Finally, Japan's decision to release the captain resolved the immediate crisis, but this had a lasting effect on regional politics.[44] Tactically, China's diplomatic and economic coercion worked and the Chinese citizen was released. However, China paid a strategic price as its maritime relations with Japan soured, creating the opportunity for a revitalised US–Japan alliance.[45] Also, since 2010, China started increasing the tempo of its activity in East China Sea, which included deployment of PLA Navy (PLAN) assets within Japan's EEZ in Miyako Strait.[46]

The true turning point in the dispute occurred in September 2012, when the Japanese government nationalised Senkaku Islands by purchasing them from their private owners. The decision by the Noda government was ostensible to pre-empt the purchase of the islands by Shintaro Ishihara, the right-wing Governor of Tokyo, so that the islands could be 'administered peacefully and stably'. To the Chinese, the purchase was a Japanese ploy to present Beijing with a *fait accompli*.[47] Beijing hit back with economic sanctions and a consumer boycott of Japanese goods, as well as an escalation of its efforts to challenge and erode Japanese control over the islands. In order to reiterate its legal claim, the Chinese government promulgated the base points and baselines of the territorial waters of the disputed islands, and their affiliated islets, and submitted it to the United Nations (UN).[48] In the military dimension, a raft of maritime and aerial incursions into the vicinity of the disputed islands occurred and has continued to increase since then (see Figure 8.1).

On 23 November 2013, China took its challenging strategy a step further by declaring an Air Defence Identification Zone (ADIZ) covering part of East China Sea and the Senkaku/Diaoyu Islands.[49] The ADIZ overlapped with the ADIZs of Japan, South Korea and Taiwan. China claimed that the setting up of the ADIZ was a 'justified act of self-defence' and not aimed at any specific country. However, the ADIZ both promoted China's claim to the Diaoyu Islands and challenged Japan's effective control. In Beijing, the imposition of the ADIZ was seen as a 'great air–sea strategic breakthrough for China'.[50] Given that China's ADIZ runs close to the strategic Miyako Strait, it would enable the PLAN to break the perceived 'encirclement' of China by the US and its allies, by going through the first island chain and into the Pacific.[51] In addition, China has rapidly intensified its activities surrounding Japan's airspace in recent years (see Figure 8.2).

Though China's regional policy seems opportunistic, its gray zone approach serves to both secure regional objectives and mobilise nationalist emotions—and at minimal cost. China's hybrid warfare approach has shown sophistication and effectiveness in dealing with the maritime disputes in the Senkaku Islands. By staying below the provocation threshold, Beijing has been able to preserve its status as a 'responsible stakeholder', while simultaneously creating psychological pressure and introducing questions about the otherwise established Japanese control of the islands. Moreover, to provoke a confrontation in which the Japanese removed Chinese personnel would defeat the objective of Beijing's current policy, which is simply to raise doubts about Japan's claim and administration, and to—in effect—put the islands in play. Beijing's longer-term objective is to force a negotiation in which Japan cedes certain rights to China to drill for and extract oil on limited plots adjacent to the Senkaku Islands.

*Gray Zone Dynamics in South China Sea*
A leading example of China's pursuit of gray zone revisionism is evident in the South China Sea. Beijing desires regional hegemony to gain control of specific resources and to counter-balance, and eventually replace, the US geopolitical pre-eminence in Asia. China's maritime assertiveness in the South China Sea, over a prolonged period, has been described as 'salami slicing', where a gradual accumulation of evidence of customary presence

Figure 8.1: Number of Intrusions of Chinese Government and Other Vessels into Japan's Territorial Sea around Senkaku

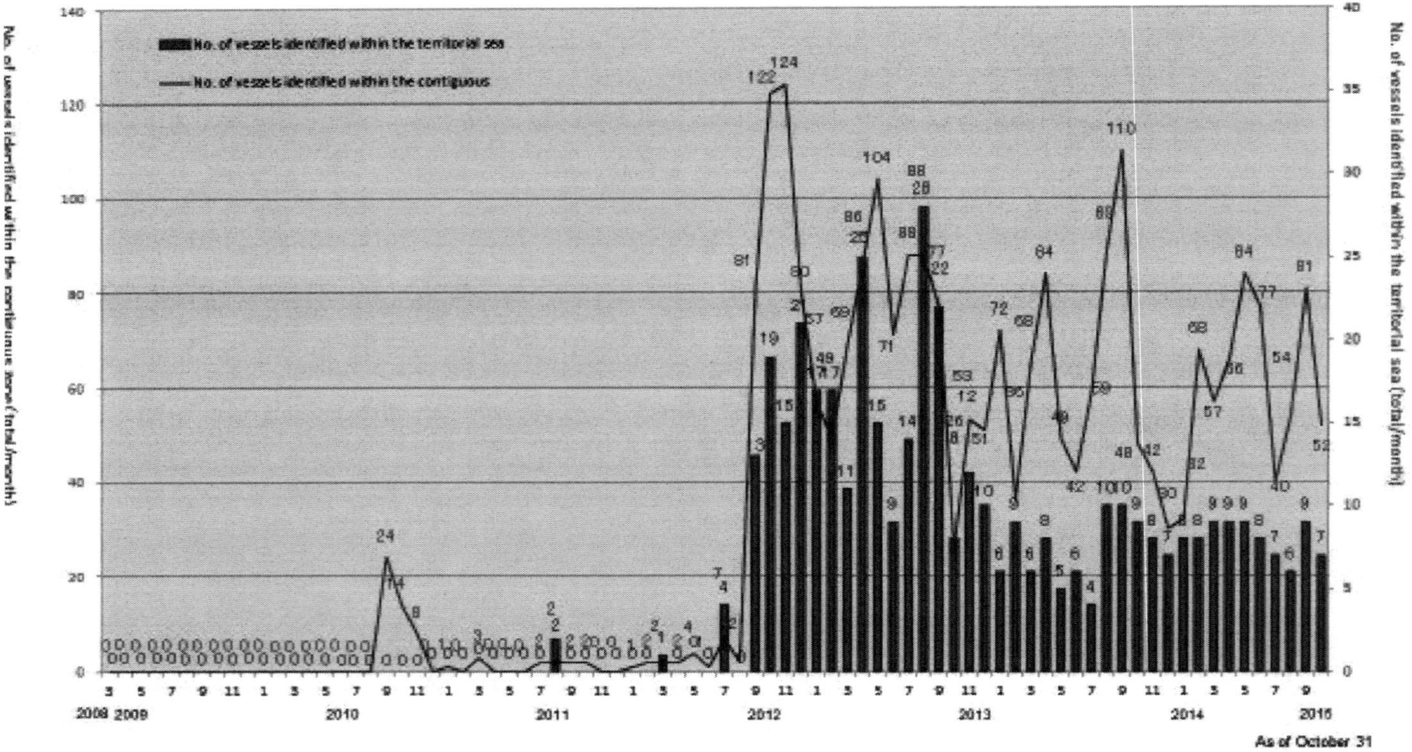

Source: Ministry of Foreign Affairs of Japan, Trends in Chinese Government and Other Vessels in the Waters Surrounding the Senkaku Islands, and Japan's Response...', available at /region/page23e_000021.html, accessed on 11 June 2017.

### Figure 8.2: Number of Scrambles by Japan against Airspace Intrusion by China

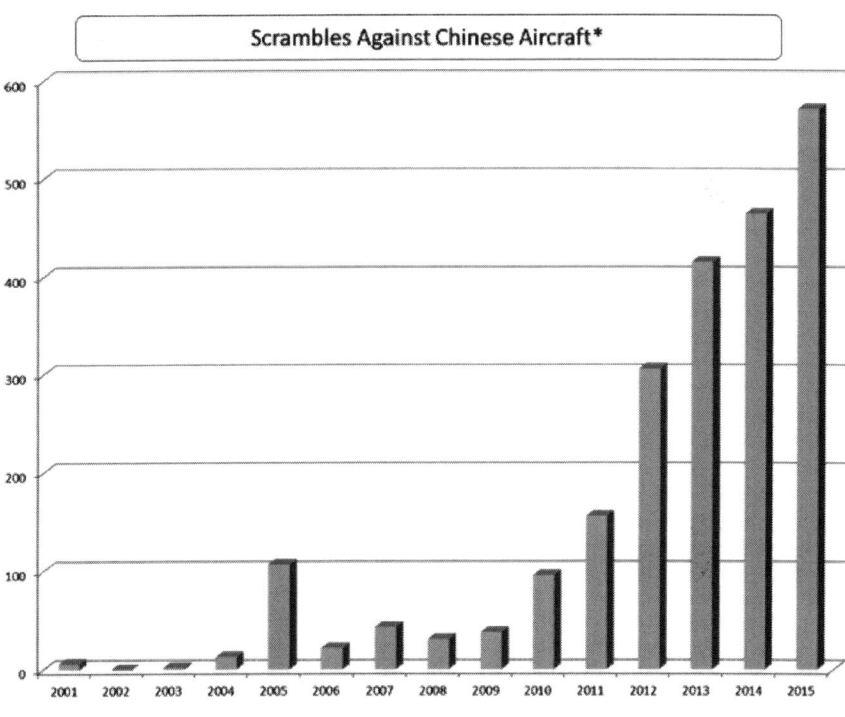

* Including presumptions

Source: Ministry of Defense of Japan, 'China's Activities Surrounding Japan's Airspace', available at http://www.mod.go.jp/e/d_act/ryouku/, accessed on 22 August 2017.

purportedly enhances China's claims to sovereignty in terms of international law and works towards an eventual settlement in its favour.[52]

To achieve its goals in South China Sea, China has taken a long series of actions to build up a persistent claim to regional hegemony—a series of steps that would appear to add up to a coherent gray zone campaign for competitive advantage. China has employed a wide range of tools and techniques as part of this campaign.[53] It has published detailed political claims to territory within its 'nine-dashed line'.[54] It has generated historical narratives and documentation in support of its claims and stated a determination to resolve disputes to its satisfaction. It has deployed a 'staggering variety and number of civil law enforcement and civilian commercial vessels and aircraft' in swarming and presence missions throughout the region; indeed, it brought together five distinct civilian maritime agencies into a unified coast guard in 2013 to enhance mutual collaboration in these forces.[55] It has employed the China National Offshore Oil Corporation for regional coercion, deploying an oil rig near the Paracel Islands in 2014. It has integrated a range of economic, diplomatic and informational steps into a coherent campaign of influence. China's approach in South China Sea reflects the geospatial thinking of a master Wei Qi player.[56] The ultimate goal is to gain control of the region. The campaign to achieve this goal relies on creeping expansion, rather than major battles.

China's creeping expansion has been in play since the last two decades. Its gray zone strategy has a number of interlocking imperatives. The first imperative is to avoid armed clashes as much as possible and initiate conflict only to exploit the favourable strategic environment. The second imperative is to control the most strategic positions in the sea through stealth if possible, or with minimum application of force. The third imperative is to strengthen own points of control through robust hubs of logistics and make them effective bases of power projection. The history of the China's involvement in South China Sea dispute has neatly followed these imperatives.[57]

In the past six decades, China has been involved in only two armed conflicts. In January 1974, China seized the western half of the Paracel Islands from Vietnam and wrested control over the Crescent Group. The second was a skirmish against unified Vietnam at Johnson South Reef in March 1988. It would be pertinent to note that both conflicts were fought at a time when there was a power vacuum in the region. During the first incident, the US was withdrawing and in the second, the Soviet Union was pulling out. The US provided tacit acquiescence in both incidents. As a result, China avoided serious diplomatic fallout.

Even though China was a latecomer in the island-grabbing spree in South China Sea, the second imperative, given earlier, is well reflected in Beijing's choice of places to occupy in the disputed areas. China occupies seven reefs as opposed to 11 by Hanoi, but five of the seven are among the most strategic features in the archipelago. Fiery Cross Reef, one of the largest in the Spratly Islands, occupies an ideal spot at the western gateway to the Spratly Islands and has the potential for land reclamation. Subi Reef, Gaven Reef, Johnson South Reef and Cuarteron Reef are located at the edge of Spratly Islands and control a large maritime area and the key waterways.[58] Mischief Reef and Scarborough Shoal, wrested by China from the Philippines through a combined subterfuge of its 'cabbage strategy'[59] along with 'small stick diplomacy',[60] provide China with a location in the northern quadrant to monitor major shipping lanes in the region. With strategic outposts in Paracel and Spratly Islands, China enjoys strategic advantage to keep South China Sea under its watchful gaze.

According to a Taipei-based newspaper, the Chinese experts have claimed in an internal PLA journal that China's massive land reclamation projects have helped the PLA to acquire a strategic advantage in military security in South China Sea to a certain extent.[61] This report confirms the earlier assessments by Asia Maritime Transparency Initiative of the Center for Strategic and International Studies that construction of military and dual-use infrastructure on the so-called big three islands in the contested Spratly chain—Subi, Mischief and Fiery Cross Reefs—is in the final stages, with the naval, air, radar and defensive facilities largely complete.[62] The satellite images offer some of the most conclusive evidence that China has continued to militarise the waters.[63] During his first state visit to the US in September 2015, Chinese President Xi Jinping had assured that 'China does not intend to pursue militarization in the Nansha/Spratly Islands.'[64] However, military installations on the South China Sea reclaimed islands seem to be an abrogation of the assurance provided. China's artificial island construction aims to strengthen its strategic position and highlights efforts towards realising its third imperative.

In South China Sea, China's 'disintegration strategy' in weakening Association of Southeast Asian Nations (ASEAN) has been clearly evident. Nothing has tested the members' institutional loyalty towards the coveted ASEAN motto of 'One Vision–One Identity–One Community' more than the South China Sea.[65] While the murmur about the lack of cohesiveness of ASEAN about the South China Sea was present even earlier, this disunity became public in June 2012, when the ASEAN Summit ended without a joint statement.[66] During diplomatic negotiations with China on the South China Sea, members of ASEAN have not been able to display a cohesive and collective approach. This disunity, again, was on full display when the ASEAN failed to issue a communique after the Hague tribunal decision.[67] It is a fact that even now, ASEAN memberstates have differing views on the issues associated with South China Sea.[68] The recent summit in Manila witnessed similar wrangling and the chairman's statement was delayed for 12 hours, with the final statement even omitting the concerns expressed in earlier years in order to avoid antagonising China.[69]

China's coercive approach through the use of its three warfares strategy has also been clearly evident in South China Sea. China has used legal warfare to achieve strategic objectives, along with psychological warfare and media warfare. It has used the framework of international law to stake a 'layered' claim in South China Sea. China's arguments unfold as follows: first, the map of the nine-dashed line is utilised as evidence of historic title over the area. Second, ancient fishing and administrative exercise by China demonstrates legal authority as concentrated over time.[70] Last, more limited forms of 'sovereignty' are claimed over the 'relevant waters' of the area for China's own purpose. China has deliberately maintained ambiguity over its jurisdictional claim, which is projected over a continuum from a maximum claim (historic waters) to a minimalist one (sovereignty): either the entire area might be Chinese internal waters (best case for Beijing) or might be subject to Chinese 'sovereignty' in some form (worst case for Beijing).[71] Since China began to build up its claim on disputed islands and reefs in the strategic waterway several years ago, Beijing also has carried out an international propaganda and influence campaign designed to sway world public opinion that its actions are not aggressive or destabilising.[72]

## Assessing Effectiveness of China's Gray Zone Strategy

It can be seen that China's gray zone strategic approach, through its assertive behaviour, has helped strengthened its claim. While it has not been able to resolve the dispute in its favour, China has been able to strengthen its position strategically, at least in the South China Sea. Gray zone concepts remain congruent with Chinese strategic thinking and leverage traditional concepts and provide a framework for modern conflict management.

China's growing claims and permanent presence in South China Sea and East China Sea exemplify effective use of 'coercive gradualism'.[73] Coercive gradualism is simply a step-by-step pursuit of one nation's interests against other nations' interests, keeping strategic tension below threshold of overt conflict. China has made gains incrementally by unilaterally changing status quo without crossing use of force red lines. None of the claimants of these maritime disputes, including China, want the full-scale war to break out. Hence, the preferred strategic approach of China has been to enhance its position through 'slow intensity conflict', which entails minor and infrequent skirmishes through small units of maritime militia in a slow and incremental escalation, avoiding sharp international focus on the Chinese violations.[74]

China's gray zone actions can be broadly differentiated into four categories: contestation of rules and norms; exploitation of rules and norms; exploitation of physical control; and contestation of physical control.[75] By contesting rules and norms, China aims to shift rules and norms in its favour through disputing the meaning of provisions and changing common practice of existing legal norms. Beijing also exploits existing rules and norms to alter the status quo in its favour, for example, promulgation of ADIZ in East China Sea and unilateral ban of fishing in the disputed water of South China Sea. China's insistence on the sovereign right to use the resources in waters that it controls and to reclaim land to construct defensive positions on features that it controls is indicative of exploitation of physical control. Examples of contestation of physical control include the incidents around the Senkaku Islands in 2010 and 2012, the standoff at Scarborough Shoal in 2012 and the harassment at Second Thomas Shoal in 2014. In all these cases, Chinese leaders sought to decisively shift the status quo by gaining control of disputed territory or maritime areas that were either uncontrolled or, in more extreme cases, controlled by an adversary.

China's playbook of gray zone coercion has broadly three elements. 'Salami slicing' or 'cabbage strategy' can be considered as the first element, which involves progressive enhancement of physical presence and gradual change of status quo. This action is simultaneously accompanied by the second element of coercive diplomatic efforts, which puts onus on the target country either to accommodate Chinese view or to risk confrontation. Third element involves astute use of three warfares strategy, which involves use of legal rhetoric to justify China's position as legitimate, along with use of media to highlights its narrative and issue threats. China has consistently demonstrated a preference for ambiguity, risk manipulation and controlling the narrative to win without fighting.[76] China's Gray Zone incidents and employments of coercive tools in the East China Sea and South China Sea has been tabulated at **Appendix (Table 8 A1)**.

As discussed earlier, gray zone strategy is congruent with Chinese strategic thinking. Specifically, it offers a structure for taking the initiative and managing uncertainty in pre-kinetic situations, although it also has applications after hostilities have begun. A formulaic approach can be seen its application of its 'three warfares', through which China makes all sovereignty claims indisputable and dismisses all counter-claims. With incremental changes to the status quo, Beijing seeks to accrue small diplomatic victories that, in addition to the tangible benefit of resource rights, may be offered up for domestic consumption as signs of Beijing's diplomatic prowess and rising global status. This simultaneously serves Beijing's economic and international agendas and its domestic political needs.

While China has managed to strengthen its strategic position in the maritime milieu, its gray zone strategy has also had some negative consequences. China's assertive approach has resulted in adverse fallouts in South China Sea and Southeast Asia, where a decade of its effective and positive diplomacy has evaporated. Even though the claimant states are maintaining a low profile, the territorial disputes are far from being resolved. China's coercion has alarmed these smaller nations, convulsed ASEAN and opened the door for an expanded US regional presence.[77] China's flagrant disregard for international law and treaty commitment has undermined Chinese credibility[78] and created a serious doubt about its role as a responsible stakeholder in a rule-based order.[79] Chinese insatiable irredentism has strengthened the balancing and hedging approach within the countries in the region.

## Coping with China's Gray Zone Strategy: Some Pointers from Doklam Standoff

It is pertinent to mention that, in addition to its maritime frontiers, China's gray zone strategy has been in play on its land frontiers as well. Brahma Chellaney has argued that China's gray zone strategy has been in play along the Indian borders. It has involved the Chinese military bringing ethnic Han pastoralists to drive Indian herdsmen from their traditional pasturelands and opening the path to salami slicing through subsequent PLA patrols.[80] While the occasional transgressions have been a source of occasional tensions, these incidents have remained lowkey and have been resolved through local flag meetings and diplomatic channels.[81]

The recent two month-long standoff at Doklam appears to be the most recent iteration of the Chinese gray zone strategy playbook as employed in the South and East China Seas. Even though the risk of a clash between India and China seemed to be over with withdrawal of troops from both sides, the standoff provides some insight into Chinese coercive playbook, along with a possible framework of mitigating strategy.

The standoff between Indian and Chinese troops at Doklam Plateau in Sikkim sector began in mid-June 2017 when China began constructing a road in Doklam Plateau in the disputed territory which forms a tri-junction between India, Bhutan and China. India and Bhutan both objected to the Chinese road construction. India intervened in the issue by sending in troops to support the Royal Bhutanese Army. This led to a halt in the construction activity by the PLA, and then began a tense standoff between India and China.[82]

The Chinese response during the standoff at Doklam was in accordance with its well-rehearsed playbook. The first element was contestation of physical control through the construction of military infrastructure in the disputed territory, along proclamation of sovereign rights over disputed territory.[83] The next element in Beijing's playbook, 'coercive diplomacy', was played out through the threat of military actions,[84] along with diplomatic efforts, to highlight that the envisaged construction was its sovereign right and India had no *locus standi* in the issue since the disputed territory was an issue between China and Bhutan.[85] The third element of three warfares was played through legal and media rhetoric.[86] Chinese legal rhetoric presented its position as legitimate and lawful through selective interpretation of historical treaties and past agreements.[87] China also leveraged its government-controlled media to highlight its narrative, signal its resolve to protect its sovereign rights and issue threats of military action.[88] Since China, in the past, has used escalating threats in the media as a precursor to China's use of force, these warnings were not a just a war of words.[89]

In this incident, India pre-empted the Chinese coercive *fait accompli* through denial—by physically denying China's bid to change the status quo—which led to a stalemate. The stalemate thwarted the preferred Chinese playbook approach of progressing consolidation of physical control under the guise of peaceful legitimate development. The onus of further escalation was, thus, shifted back to China. Notwithstanding the ever-present risk of escalation of standoff into overt conflict, India demonstrated its resolve in preserving its defined red line. India also managed to keep its alliance with Bhutan intact, despite the Chinese efforts to cleave Bhutan away through inducement.[90] In fact, India, in spite of provocation from China, behaved like a mature great power through its firm and restrained approach.

After a tense standoff lasting nearly two months, the denouement saw a mutual withdrawal by both sides and removal of construction equipment by China. The immediate risks of conflict have receded, but the border dispute remains unresolved. Even though China seems to have backed down, Chinese statements suggest that it has not changed its position on the border tri-junction.

While it is too early to draw definitive lessons from this standoff, the following points are relevant for a mitigating strategy against the Chinese gray zone approach:[91]

- Chinese behaviour in territorial disputes is more likely to be deterred by denial than by threats of punishment. China will continue the combination of consolidating its physical presence

and engaging in coercive diplomacy, lawfare and media campaigns, unless it is stopped directly.
- Denial strategies may be effective, but they have their limitations. Denial is inherently risky. Countering China's playbook involves risk of escalation.
- The agreement to disengage suggests that Beijing's position in crises can be flexible, and perhaps responsive to assertive counter-coercion.

## Future Contours of China's Gray Zone Strategy

In the coming decades, China will continue to play the coordinated 'gray zone' strategic game through refinement of its non-kinetic tools to protect and advance its long term interest. China will leverage its accumulated experience in managing gray zone incidents and use the three warfares as an instrument of first resort in cases where the direct military action is likely to have negative collateral effects, for example, in the diplomatic and opinion realms. In order to shape international opinion, China will continue to strengthen its international propaganda arms, namely, *Xinhua* and *CCTV*, to enhance its image and soft power. A clear focus on a network for perception management, mentioned in the white paper on strengthening maritime cooperation in the Maritime Silk Route, is a pointer in that direction.[92] China will also be more aggressive, and perhaps influential, in the interpretation and development of international law, especially regarding United Nations Convention on the Law of the Sea (UNCLOS).[93] China will continue to target the historical foundations of international law and will advance a Chinese-friendly historical narrative that gives weight to China's contemporary arguments. Maritime security agencies will remain on the 'front line' to manage incidents. They will assume the leading role, with the PLAN in the background. China will continue to use the 'small stick' of law enforcement and civilian assets, including fishing vessels, to wear out small Asian navies and coast guards. Beijing will be able to thus assert control over larger areas further afield in an administrative or law enforcement capacity.[94]

In essence, China now has a playbook for resource acquisition and conflict management that can be tailored to address each new situation and targets even beyond its maritime periphery. The future key arena of the Chinese gray zone strategic application is likely to be the Indian Ocean. While anti-piracy has provided a plausible rationale for the deployment of PLAN assets, China has been progressively strengthening its presence in the Indian Ocean region through the deployment of submarines, research vessels and intelligence-gathering vessels. China has also acquired controlling stakes in various ports, namely, Kyaukpyu (Myanmar), Hambantota (Sri Lanka) and Gwadar (Pakistan).

From the Indian perspective, the next decade will be a critical juncture in the Indo-Pacific in general, and India–China relation in particular. Given the truism of 'the past as a prologue', China's gray zone strategy, and its tools of 'three warfares', is likely to intensify in the Indian Ocean region, in light of its expanding strategic and economic interest, through the Belt and Road Initiative and Maritime Silk Route.

## NOTES

1. Sanjeev Miglani, 'China Is a Disruptive Force, U.S. Pacific Military Chief Says', *Reuters*, 18 January 2018, https://www.reuters.com/article/us-india-security-china-us/china-is-a-disruptive-force-u-s-pacific-military-chief-says-idUSKBN1F71ZS, accessed on 20 Jan 2018.
2. James Holmes and Toshi Yoshihara, 'Command of the Sea with Chinese Characteristics', *Orbis*, Vol. 49, No. 4, 2005, p. 681.
3. Michael Green, Kathleen Hicks, Zack Cooper, John Schaus and Jake Douglas, *Countering Coercion in Asia: Theory and Practice of Grey Zone Deterrence*, Washington, DC: Center for Strategic & International Studies, May 2017, pp. 3–13, available at https://csis-prod.s3.amazonaws.com/s3fs-public/publication/170505_Green M_CounteringCoercionAsia_Web.pdf?OnoJXfWb4A5gw_n6G.8azgEd8zRIM4wq, accessed on 20 July 2017.
4. Hal Brands, 'Paradoxes of the Gray Zone', Foreign Policy Research Institute, 5 February 2016, available at http://www.fpri.org/article/2016/02/paradoxes-gray-zone/, accessed on 9 June 2017.
5. Michael J. Mazarr, *Mastering the Gray Zone: Understanding a Changing Era of Conflict*, Carlisle, PA: Strategic Studies Institute, Army War College Press, 2015, p. 82.
6. Frank Hoffman, 'The Contemporary Spectrum of Conflict: What is War?', in *2016 Index of U.S. Military Strength*, available at http://index.heritage.org/military/2016/essays/contemporary-spectrum-of-conflict/, accessed on 7 July 2017.
7. Henry Kissinger, *On China*, London: Penguin Books Ltd, 2012, p. 47.
8. Ibid.
9. David Lai, *Learning from the Stones: A Go Approach to Mastering China's Strategic Concept*, Shi, Advancing Strategic Thought Series, Carlisle, PA: Strategic Studies Institute, Army War College Press, 2004, pp. 6–8.
10. Ibid.
11. David Lai, 'China's Strategic Moves and Counter-moves', *Parameters*, Vol. 44, No. 4, 2014, p. 13.
12. Benjamin David Baker, 'Hybrid Warfare with Chinese Characteristics', *The Diplomat*, 23 September 2015, available at http://thediplomat.com/2015/09/hybrid-warfare-with-chinese-characteristics/, accessed on 26 July 2017.
13. Tiejun Zhang, 'Chinese Strategic Culture: Traditional and Present Features', *Comparative Strategy*, Vol. 21, No. 2, 2002, pp. 73–90.

14. Ralph D. Sawyer and Mei-chün Sawyer (eds), *The Seven Military Classics of Ancient China* [*Wu Jing Qi Shu*], Boulder: Westview Press, 1993, p. 5.
15. Timothy L. Thomas, 'Geothinking Like the Chinese: A Potential Explanation of China's Geostrategy', Foreign Military Studies Office, Fort Leavenworth, Kansas, 2009, pp 15–18, available at http://www.usafa.edu/app/uploads/Thomas-Geothinking-Like-the-Chinese-2011.pdf, accessed on 14 April 2017.
16. Secretary of Defense, *Annual Report on the Military Power of the People's Republic of China FY 2002*, Washington, DC: US Department of Defense, 2002, pp. 5–6, available at http://www.globalsecurity.org/military/library/report/2002/d20020712china.pdf, accessed on 14 April 2017.
17. Thomas, *Geothinking Like the Chinese*, p. 17.
18. Lai, *Learning from the Stones*, n. 8, p. 2.
19. Michael Pillsbury, *The Hundred-Year Marathon: China's Secret Strategy to Replace America as the Global Superpower* (1st edition), New York: Henry Holt and Company, 2015.
20. Sun Tzu, *The Art of War*, translated by Samuel B. Griffith, Oxford: Oxford University Press, 1980, p. 66.
21. Thomas G. Mahnken, *Secrecy and Stratagem: Understanding Chinese Strategic Culture*, Woollahra, NSW: Lowy Institute for International Policy, 2011, p. 21.
22. Peng Guangqian and Yao Youzhi (eds), *The Science of Military Strategy*, Beijing: Military Science Publishing House, 2005, p. 31. See also ibid., pp. 72–77.
23. Qiao Liang and Wang Xiangsui, *Unrestricted Warfare*, translated by Foreign Broadcast Information Service, US, Beijing: PLA Literature and Arts Publishing House, 1999, available at https://www.cryptome.org/cuw.htm, accessed on 18 May 2017
24. Tony Corn, 'Peaceful Rise through Unrestricted Warfare: Grand Strategy with Chinese Characteristics', *Small Wars Journal*, 5 January 2010, available at http://smallwarsjournal.com/journal/iss/v6n6.pdf, accessed on 1 June 2017.
25. Dean Cheng, '*Unrestricted Warfare*: Review Essay II', *Small Wars & Insurgencies*, Vol. 11, No. 1, 2000, pp. 122–123.
26. Sangkuk Lee, 'China's "Three Warfares": Origins, Applications, and Organizations', *Journal of Strategic Studies*, Vol. 37, No. 2, 2014, pp. 198–221.
27. Ibid, p 202.
28. Office of the Secretary of Defense, *Annual Report to Congress: Military and Security Developments Involving the People's Republic of China 2011*, Washington DC: Department of Defense 2011, p. 26, available at https://www.defense.gov/Portals/1/Documents/pubs/2011_CMPR_Final.pdf, accessed 17 May 2017. Also see Emilio Lasiello, 'China's Three Warfares Strategy Mitigates Fallout from Cyber Espionage Activities', *Journal of Strategic Security*, Vol. 9, No. 2, 2016, pp. 47–71, doi:10.5038/1944-0472.9.2.1489 (Accessed May 13, 2017).
29. Information Office of the State Council of the People's Republic of China, 'China's National Defense in 2006', December 2006, available at http://www.chinadaily.com.cn/china/2006-12/29/content_5425025_3.htm, accessed on 12 June 2017.
30. Laura Jackson, 'Revisions of Reality: The Three Warfares—China's New Way of War', in *Information at War: From China's Three Warfares to NATO's Narratives*, London: Legatum Institute, 2015, pp. 5–13, available at https://lif.blob.core.windows.net/lif/docs/default-source/publications/information-at-

war-from-china-s-three-warfares-to-nato-s-narratives-pdf.pdf?sfvrsn=2, accessed on 6 June 2017.
31. Stefan Halper (ed.), *China: The Three Warfare*, Washington, DC: Office of Net Assessment, Office of the Secretary of Defense, 2013, p. 12, available at http://cryptomeorg.siteprotect.net/prc-three-wars.pdf, accessed on 13 March 2017; emphasis in original.
32. Dean Cheng, 'Winning without Fighting: Chinese Public Opinion Warfare and the Need for a Robust American Response', The Heritage Foundation, 2012, available at http://www.heritage.org/Home/Research/Reports/2012/11/Winning%20Without%20Fighting%20Chinese%20Public%20Opinion%20Warfare%20and%20the%20Need%20for%20a%20Robust%20American%20Response, accessed on 6 March 2017.
33. Halper (ed.), *China: The Three Warfares*, n. 29, p. 13.
34. Timothy A. Walton, 'Treble Spyglass, Treble Spear: China's "Three Warfares'. Defense Concepts. Volume 4, Edition 4, 2009, p. 51; see also Halper, 'China: Three Warfares', p. 30.
35. Elsa Kania, 'The PLA's Latest Strategic Thinking on the Three Warfares', *China Brief*, Vol. 16, No. 3, 2016, available at https://jamestown.org/program/the-plas-latest-strategic-thinking-on-the-three-warfares/, accessed on 26 June 2017.
36. 'China: Toward a Less Cooperative, More Assertive Posture', in Yoshiaki Sakaguchi (ed.), *East Asian Strategic Review 2011*, Tokyo: National Institute for Defense Studies, 2011, p. 131, available at http://www.nids.mod.go.jp/english/publication/east-asian/pdf/2011/east-asian_e2011_04.pdf, accessed on 13 March 2017.
37. Fumio Ota, 'Sun Tzu in Contemporary Chinese Strategy', *Joint Force Quarterly*, Vol. 73, No. 2, 2014, pp. 76–80.
38. 'What to Do about Tensions in Asia', Council on Foreign Relations, 28 January 2014, available at https://www.cfr.org/event/what-do-about-tensions-asia-0, accessed on 4 April 2017.
39. James R. Holmes and Toshi Yoshihara, 'Deterring China in the "Gray Zone": Lessons of the South China Sea for U.S. Alliances', *Orbis*, Vol. 61, No. 3, 2017, pp. 322–339.
40. M. Taylor Fravel, 'Explaining Stability in the Senkaku (Diaoyu) Islands Dispute', in Gerald Curtis, Ryosei Kokubun and Wang Jisi (eds), *Getting the Triangle Straight: Managing China–Japan–US Relations*, Washington, DC: Brookings Institution Press, 2010, pp. 144–164.
41. William Choong, 'The Senkaku/Diaoyu Dispute', *Adelphi Series*, Vol. 54, No. 445, 2014, pp. 59–92.
42. Ibid., p. 77.
43. Jeff Himmelman and Ashley Gilbertson, 'A Game of Shark and Minnow', *The New York Times*, 24 October 2013, available at http://www.nytimes.com/newsgraphics/2013/10/27/south-china-sea/, accessed on 17 May 2017.
44. David Fouse, 'Japan's 2010 National Defense Program Guidelines: Coping with the Grey Zones', DTIC Document, 2011, available at http://oai.dtic.mil/oai/oai?verb=getRecord&metadataPrefix=html&identifier=ADA560105, accessed on 17 May 2017.
45. Green et al., 'Countering Coercion in Asia', n. 2, p. 94.

46. 'China: Toward a Less Cooperative, More Assertive Posture', n. 34, pp. 134–136.
47. Sanaa Yasmin Hafeez, 'The Senkaku/Diaoyu Islands Crises of 2004, 2010, and 2012: A Study of Japanese–Chinese Crisis Management', *Asia-Pacific Review*, Vol. 22, No. 1, 2015, pp. 84–85.
48. International Crisis Group, 'Dangerous Waters: China–Japan Relations on the Rocks', Asia Report No. 245, 8 April 2013, available at https://d2071andvip0wj.cloudfront.net/dangerous-waters-china-japan-relations-on-the-rocks.pdf, accessed on 19 March 2017.
49. Ian E. Rinehart and Bart Elias, 'China's Air Defense Identification Zone (ADIZ)', Congressional Research Service Report, Washington, DC, 30 January 2015, available at https://fas.org/sgp/crs/row/R43894.pdf, accessed on 2 June 2017.
50. Harry White, 'The ADIZ and Rebalancing on the Run', *The Strategist*, 28 November 2013, available at https://www.aspistrategist.org.au/the-adiz-and-rebalancing-on-the-run/, accessed on 1 Jun 2017.
51. Choong, 'The Senkaku/Diaoyu Dispute', n. 39, pp. 83–84.
52. Ronald O'Rourke, 'Maritime Territorial and Exclusive Economic Zone (EEZ) Disputes Involving China: Issues for Congress', DTIC Document, 2015, p. 23, available at http://oai.dtic.mil/oai/oai?verb=getRecord&metadataPrefix=html&identifier=ADA624220, accessed on 28 December 2016.
53. Mohan Malik, 'Historical Fiction: China's South China Sea Claims', *World Affairs*, May–June 2013, available at www.worldaffairsjournal.org/article/historical-fiction-china%E2%80%99s-south-china-sea-claims, accessed on 2 January 2017.
54. US Department of State, Bureau of Oceans and International Environmental and Scientific Affairs, 'Limits in the Seas, China: Maritime Claims in the South China Sea', No. 143, 5 December 2014, available at https://www.state.gov/documents/organization/234936.pdf, accessed on 7 January 2017.
55. James Kraska, 'How China Exploits a Loophole in International Law in Pursuit of Hegemony in East Asia', Foreign Policy Research Institute, 22 January 2015, available at https://www.fpri.org/article/2015/01/how-china-exploits-a-loophole-in-international-law-in-pursuit-of-hegemony-in-east-asia/, accessed on 4 April 2017.
56. James Holmes, 'Maritime Southeast Asia: A Game of Go?', *The Diplomat*, 20 December 2014, available at https://www.evernote.com/Home.action#n=bd39b6d5-963e-4b3f-acde-9f98ce8e2cc8&ses=4&sh=2&sds=5&, accessed 6 February 2017.
57. Alexander L. Vuving, 'China's Grand-Strategy Challenge: Creating its Own Islands in the South China Sea', *The National Interest*, 8 December 2014, available at http://nationalinterest.org/feature/chinas-grand-strategy-challenge-creating-its-own-islands-the-11807?page=show, accessed on 6 February 2017.
58. Chris Rahman and Martin Tsamenyi, 'A Strategic Perspective on Security and Naval Issues in the South China Sea', *Ocean Development & International Law*, Vol. 41, No. 4, 2010, p. 319.
59. Cheney-Peters Scott, 'A Feast of Cabbage and Salami: Part I—The Vocabulary of Asian Maritime Disputes', Center for International Maritime Security, 29 October 2014, available at http://cimsec.org/feast-cabbage-salami-part-vocabulary-asian-maritime-disputes/13441, accessed on 17 February 2017.
60. James Holmes, 'Small-Stick Diplomacy in the South China Sea', *The National Interest*, 23 April 2012, available at http://nationalinterest.org/commentary/

small-stick-diplomacy-the-south-china-sea-6831?page=show, accessed on 17 February 2017.

61. 'Chinese's Military Dominance in S. China Sea Complete: Report', *ABS-CBN News*, 20 March 2017, available at http://news.abs-cbn.com/overseas/03/20/17/chineses-military-dominance-in-s-china-sea-complete-report, accessed on 14 May 2017.

62. Gregory B. Poling, 'Prepare for a Stormy 2017 in the South China Sea', Asia Maritime Transparency Initiative, 12 January 2017, available at https://amti.csis.org/prepare-stormy-2017-south-china-sea/, accessed on 13 January 2017.

63. 'China Archives', Asia Maritime Transparency Initiative, available at https://amti.csis.org/island-tracker/chinese-occupied-features/, accessed on 29 May 2017.

64. Jeremy Au Yong, 'Tensions Remain as Xi's US Visit Ends', *The Straits Times*, 27 September 2015, available at http://www.straitstimes.com/world/united-states/tensions-remain-as-xis-us-visit-ends, accessed on 30 May 2017.

65. Jimmy Chalk and Sarah Grant 'Water Wars: ASEAN Soft-pedals South China Sea Disputes', *Lawfare*, 5 May 2017, available at https://www.lawfareblog.com/water-wars-asean-soft-pedals-south-china-sea-disputes, accessed on 16 May 2017.

66. David Tweed and David Roman, 'Chinese–ASEAN Meeting on South China Sea Ends in Confusion', *The Japan Times*, 15 June 2016, available at http://www.japantimes.co.jp/news/2016/06/15/asia-pacific/chinese-asean-meeting-south-china-sea-ends-confusion/, accessed on 23 January 2017.

67. Manuel Mogato, Michael Martina and Ben Blanchard, 'ASEAN Deadlocked on South China Sea, Cambodia Blocks Statement', *Reuters*, 26 July 2017, available at http://www.reuters.com/article/us-southchinasea-ruling-asean-idUSKCN1050F6, accessed on 28 July 2017.

68. Liu Zhen and Catherine Wong, 'Most Asean Countries 'want to Stay out of Beijing's South China Sea Dispute with the Philippines'", *South China Morning Post*, 28 July 2016, available at http://www.scmp.com/news/china/diplomacy-defence/article/1995991/most-asean-countries-want-stay-out-beijings-south-china, accessed on 1 August 2017.

69. Manuel Mogato, 'ASEAN Gives Beijing a Pass on South China Sea Dispute, cites "Improving Cooperation"', *Reuters*, 30 April 2017, available at http://www.reuters.com/article/us-asean-summit-idUSKBN17W02E, accessed on 7 May 2017.

70. Zhiguo Gao and Bing Bing Jia, 'The Nine-Dash Line in the South China Sea: History, Status, and Implications', *The American Journal of International Law*, Vol. 107, No. 1, 2013, pp. 98–124.

71. Gregory B. Poling, 'Time to End Strategic Ambiguity in the South China Sea', Center for Strategic and International Studies, 6 July 2012, available at https://www.csis.org/analysis/time-end-strategic-ambiguity-south-china-sea, accessed on 23 December 2016.

72. Bill Gertz, 'China's Presses the Fast Forward Button to Dominate the South China Sea', *The National Interest*, 1 June 2016, available at http://nationalinterest.org/blog/the-buzz/chinas-presses-the-fast-forward-button-dominate-the-south-16416, accessed on 22 August 2017.

73. William G. Pierce, Douglas G. Douds and Michael A. Marra, 'Understanding Coercive Gradualism', *Parameters*, Vol. 45, No. 3, 2015, p. 51.
74. Andrew Scobell, 'Slow-intensity Conflict in the South China Sea', Foreign Policy Research Institute, 13 August 2000, available at http://www.fpri.org/article/2000/08/slow-intensity-conflict-in-the-south-china-sea/, accessed on 22 December 2017.
75. Green et al., 'Countering Coercion in Asia', n. 2, pp. 266–267.
76. Oriana Skylar Mastro and Arzan Tarapore, 'Countering Chinese Coercion: The Case of Doklam', *War on the Rocks*, 29 August 2017, available at https://warontherocks.com/2017/08/countering-chinese-coercion-the-case-of-doklam/, accessed on 1 September 2017.
77. Halper (ed.), *China: The Three Warfares*, n. 29.
78. Namrata Goswami, 'Can China be taken Seriously on its "Word" to Negotiate Disputed Territory?', *The Diplomat*, 18 August 2017, available at http://thediplomat.com/2017/08/can-china-be-taken-seriously-on-its-word-to-negotiate-disputed-territory/, accessed on 23 August 2017.
79. Koh Swee Lean Collin, 'America: China doesn't Care about Your Rules-based Order', *The National Interest*, 17 August 2017, available at http://nationalinterest.org/feature/america-china-doesnt-care-about-your-rules-based-order-21943 accessed on 23 August 2017.
80. Brahma Chellaney, 'China's Salami-slice Strategy', *The Japan Times*, 25 July 2013, available at http://www.japantimes.co.jp/opinion/2013/07/25/commentary/world-commentary/chinas-salami-slice-strategy/, accessed on 14 December 2016.
81. Ministry of External Affairs, Government of Indian, 'Lok Sabha Question No.4462: Relations with China', 14 December 2016, available at http://www.mea.gov.in/lok-sabha.htm?dtl/27829/QUESTION_NO4462_RELATIONS_WITH_CHINA, accessed on 26 September 2017.
82. 'From Doklam Standoff to Disengagement: How India and China Resolved the Crisis', *The Indian Express*, 31 August 2017, available at http://indianexpress.com/article/india/doklam-india-china-bhutan-sikkim-narendra-modi-4822205/, accessed on 2 September 2017.
83. 'Doklam has Always been under China's Effective Jurisdiction without Dispute: FM', *Xinhua*, 24 July 2017, available at http://news.xinhuanet.com/english/2017-07/24/c_136468821.htm, accessed on 26 September 2017.
84. 'Chinese Envoy Calls for Unconditional Withdrawal of Indian Border Troops, Peaceful Solution to Incident in Doklam', *Xinhua*, available at http://news.xinhuanet.com/english/2017-07/06/c_136421698.htm, accessed on 26 September 2017.
85. 'Full Text of Facts and China's Position Concerning Indian Border Troops' Crossing of China–India boundary', *China Daily*, 3 August 2017, available at http://www.chinadaily.com.cn/world/2017-08/03/content_30341027.htm, accessed on 26 September 2017. See also Ministry of Foreign Affairs of China, 'Foreign Ministry Spokesperson Geng Shuang's Remarks on the Indian Border Troops' Illegal Crossing of the China–India Boundary into the Chinese Territory', 3 August 2017, available at http://www.fmprc.gov.cn/mfa_eng/xwfw_665399/s2510_665401/t1482345.shtml, accessed on 26 September 2017.
86. Indrani Bagchi, 'Doklam Standoff: China Playing out its 'Three Warfares' Strategy

against India', *The Times of India*, 13 August 2017, available at http://timesofindia.indiatimes.com/india/china-playing-out-its-three-warfares-strategy-against-india/articleshow/60036197.cms, accessed on 26 September 2017.

87. 'Infographic: Under the Pretext of "Protecting Bhutan", India Attempts to Create Disputes in Doklam', *Xinhua*, 3 August 2017, available at http://news.xinhuanet.com/english/2017-08/03/c_136497278.htm, accessed on 26 September 2017.

88. 'India Repeats Mistakes of 1962 with Continued Violation of China's Sovereignty: Former Journalist', *Global Times*, 26 July 2017, available at a http://www.globaltimes.cn/content/1058095.shtml, accessed 26 September 2017. See also 'China Never Backs down in Defending Sovereignty', *Global Times*, 17 August 2017, available at http://www.globaltimes.cn/content/1061770.shtml, accessed on 6 September 2017; 'Commentary: Turning a Deaf Ear to China will Not Help India on Doklam', *Xinhua*, 14 July 2017, available at http://news.xinhuanet.com/english/2017-07/14/c_136444504.htm, accessed on 26 September 2017; and 'Modi mustn't Pull India into Reckless Conflict', *Global Times*, 4 August 2017, available at http://www.globaltimes.cn/content/1059715.shtml, accessed on 26 September 2017.

89. Paul H.B. Godwin and Alice L. Miller, *China's Forbearance has Limits: Chinese Threat and Retaliation Signaling and its Implications for a Sino-American Military Confrontation*, Washington, DC: National Defense University Press, April 2013, available at http://www.dragonreport.com/Dragon_Report/home/home_files/ChinaPerspectives-6.pdf, accessed on 26 September 2017.

90. Yujo Kuronuma, 'China Woos Bhutan with $10 billion in Standoff with India', *Nikkei Asian Review*, 24 August 2017, available at https://asia.nikkei.com/Spotlight/China-India-Tensions/China-woos-Bhutan-with-10-billion-in-standoff-with-India, accessed on 26 September 2017.

91. Mastro and Tarapore, 'Countering Chinese Coercion: The Case of Doklam', n. 74.

92. Abhay K. Singh, 'Unpacking China's White Paper on Maritime Cooperation under BRI', *IDSA Issue Briefs*, 28 June 2017, available at http://www.idsa.in/issuebrief/chinas-white-paper-on-maritime-cooperation-under-bri_aksingh_280617, accessed on 23 August 2017.

93. James Goldrick, 'The Three Warfares in the Maritime Domain: A Look Ahead', in Halper (ed.), *China: The Three Warfares*, n. 29.

94. Peter Mattis, 'Chinese Propaganda and Positioning in the Sino-American Crises: The EP-3 and the Impeccable Cases', in Halper (ed.), *China: The Three Warfares*, n. 29.

# APPENDIX

**Table 8A.1: Gray Zone Incidents in East China Sea and South China Sea**

| Sl. No. | Incident | Category | Gray Zone Strategy | Implementation | Status |
|---|---|---|---|---|---|
| 1. | **Senkaku Island Trawler Collision** On 7 September 2010, a Chinese fishing trawler collided with JCG vessels in waters near the disputed Senkaku Islands. Unlike in past incidents, Tokyo decided to arrest the skipper and detain his crew, triggering a two-week diplomatic crisis. | **Contestation of Physical Control** | Cabbage Strategy | Enhancement of presence. | Case against Chinese trawler withdrawn and Japan released the Chinese boat and crew. China continued to maintain presence around the disputed islands. East China Sea oil and gas negotiations not resumed. |
| | | | Diplomacy | Demands unconditional release of trawler and crew. Cancels parliamentary exchange. Suspends bilateral exchanges. Fujita Corporation employees arrested. | |
| | | | Economy | Suspends East China Sea oil and gas negotiations. Tourist boycott. Embargo on rare earth export. | |
| | | | Military | Starts Senkaku patrol. | |
| | | | Three Warfares | Media highlights 'Japanese harassment' of Chinese fishing boats. China's legal rights over Senkaku Islands emphasised. Demonstration of public anger through protests. Chinese officials blame Japan for the incident. | |

| | | Contestation of Physical Control | Cabbage Strategy | |
|---|---|---|---|---|
| 2. | **Scarborough Shoal Standoff (2012)** On 10 April 2012, a Philippine warship intercepted several Chinese fishermen at the disputed Scarborough Shoal, leading to a two-month standoff when two Chinese law enforcement vessels arrived before Manila could complete the arrest. Despite initial de-escalations, the two sides could not agree on the terms of a total withdrawal. Manila then announced that it would seek international arbitration as well as help from ASEAN and the US. | | Progressive enhancement of presence. | |
| | | | Diplomacy | Denounces violation of Chinese sovereignty. Protests presence of the Philippines Navy. Warns escalation if outsider's (the US) support sought. Persuades ASEAN against any statement on the standoff. Warns against effort to internationalise dispute through arbitration. | After diplomatic negotiations brokered by the US, both side agreed on mutual withdrawal. Reports differ on whether China actually agreed to a final mutual withdrawal, under what terms and through which diplomatic channel. Yet, on 15 June, the Philippines' vessels left the shoal while China's either remained or quickly returned and began denying entry to Filipino fishermen, resulting in a *de facto* seizure of control by Beijing. |
| | | | Economy | Imposes quarantine on fruit imports from the Philippines. | |
| | | | Military | Threatens escalation and enhances presence of Chinese government vessels. Commences air patrols by military aircraft. Ratchets up tensions at sea. | The Philippines initiated arbitration proceedings which was boycotted by China. The Permanent Court of Arbitration, on 12 July 2016, ruled in favour of the Philippines. Arbitration ruling was not accepted by China. Manila decided to pursue the issue bilaterally with China and chose not to seek enforcement of the court ruling. Scarborough Shoal remains under *de facto* control of China. |
| | | | Three Warfares | Media narratives highlight harassment of China's fishermen as a violation of Chinese sovereignty. Releases a position paper on Chinese claims to Scarborough Shoal emphasising full ownership of the shoal. Nationalist sentiments provoked against the Philippines. Chinese hackers defaced the Philippines university website. | |

*Expanding the Turbulent Maritime Periphery* 169

170  Hybrid Warfare: The Changing Character of Conflict

| Sl. No. | Incident | Category | Gray Zone Strategy | Implementation | Status |
|---|---|---|---|---|---|
| 3. | **East China Sea ADIZ (2013)** On 23 November 2013, Beijing announced its first ADIZ to better monitor and control international airspace in much of the East China Sea. The US, Japan and other regional states quickly criticised the decision, particularly for its perceived coercive intent; the application of rules to foreign aircraft transiting the zone but not entering Chinese national airspace; inclusion of airspace above disputed territory; overlap with the existing ADIZs of other states; and threats of 'emergency defensive measures' against non-compliance. | **Exploitation of Rule and Norms** | Strengthening claim over claimed EEZ and continental shelf | Map of EEZ congruent with geographical coordinate of claimed EEZ and continental shelf included disputed territory. | Despite overwhelming diplomatic opposition, Beijing refused to redraw its promulgated ADIZ. On the other hand, most countries ultimately chose to accommodate China's requirements for commercial airlines. |
| | | | Diplomacy | Plan discussed with Japan and South Korea but disregards objections. ADIZ implemented unilaterally with one-hour notice. Justified the decision on the basis of China's 1995 Law on Civil Aviation, 1997 Law on National Defense and 2001 Basic Rules on Flight. Insisted China's ADIZ consistent with International Civil Aviation Organization (ICAO) norms and common international practice. | |
| | | | Military | Warns adoption of defensive emergency measures to respond to aircraft that do not cooperate in the identification or refuse to follow the instructions. Deploys early warning aircraft and fighters for ADIZ patrol. | |

| Sl. No. | Incident | Category | Gray Zone Strategy | Implementation | Status |
|---|---|---|---|---|---|
| | | | Three Warfares | Media narrative highlighted that China established an ADIZ to better 'guard against potential air threats' and help 'defend' its sovereign airspace. Chinese experts argued legal validity of Chinese ADIZ, its congruence with ICAO norms and common international practices. | |
| 4. | **Spratly Islands Land Reclamation (2013–)** In late 2013, China launched a massive campaign of land reclamation and construction on the seven reefs it already occupied in the Spratly Islands. Dredgers, barges and work crews began expanding four Chinese outposts between December 2013 and May 2014. | **Exploitation of Physical Control** | Salami Slicing | Limited reclamation and construction by all claimants. Scale and speed of Chinese reclamation efforts have been astounding. Reclamation efforts were accelerated post the commencement of arbitral proceedings by the Philippines in January 2013. | Till late 2015, China added over 3,200 acres of land to the seven features (Mischief Reef, Cuarteron Reef, Fiery Cross Reef, Gaven Reef [North], Johnson Reef, Hughes Reef and Subi Reef) it occupies; other claimants reclaimed approximately 50 acres of land over the same period. China, in effect, built large artificial islands with airstrips, port facilities, buildings and other installations, including military equipment. |
| | | | Diplomacy | Highlights land reclamation activity within the scope of China's sovereignty. Rejects moratorium proposal on land reclamation by the Philippines. Diplomatic engagement with ASEAN to avoid adversarial reference. | |

| Sl. No. | Incident | Category | Gray Zone Strategy | Implementation | Status |
|---|---|---|---|---|---|
| | | | Military | Naval escorts to dredger employed for land exclamation. Placement of surveillance assets for defensive military equipment of reclaimed islands. | |
| | | | Three Warfares | Propagates narrative about reclamation activity within sovereign rights of China. Chinese reclamations efforts will enhance search and rescue efforts in South China Sea. | |

# 9

# India and Hybrid Warfare

*Alok Deb*

The liberation of Bangladesh by the Indian Armed Forces was a classic example of strategic manoeuvre, resulting in the surrender of 93,000 men, surpassing even the capitulation of Field Marshal von Paulus's 6th Army at Stalingrad in terms of the number of prisoners of war (PoWs), assets and territory captured. Even so, a common criticism about the Indian Army over the years has been its propensity to rigidly adhere to a traditional mindset. Teaching and practice of operational art and strategy are still influenced by conventional scenarios where victory or defeat is measured in the tangibles of earlier eras. Despite the sister services having wider perspectives on the matter, overall for the Indian military, the requirement for incorporating new concepts and doctrines which conform to current and evolving realities of warfare in the information age, and translating these into strategy, continues to be an ongoing concern. Extrapolating further, the major challenge for statesmen, soldiers and others concerned with national security in the uncertain backdrop of the twenty-first century geopolitics in the Indo-Pacific is to ensure that India's interests are never compromised, and evolve concepts, doctrines and strategies guaranteeing the same. Both these requirements mandate fashioning of synergised instruments of national power, with the capability to protect the national interest in any contingency, while retaining effective and publicised capacities for deterrence.

The traditional roles of India's military remain confined to defence of the borders, combating internal disorder and providing aid to civil authorities in times of distress, when called upon to do so. With India's heightened profile in the region, some of these roles have expanded (humanitarian assistance and disaster relief), while newer additions such as catering for out-of-area contingencies are in the process of being formalised within the services. Efforts at capability building to fulfil such roles are ongoing. Most such endeavours, however, are templated on concepts of 'state-on-state warfare', with limited efforts made, till date, towards gearing up for future challenges of the hybrid kind. As highlighted in various chapters of this study, while asymmetric, guerrilla or unconventional warfare, along with use of ruses and other stratagems, have been a part of warfighting from time immemorial, it is the insertion of ever-evolving technology and the instantaneous dissemination of information across boundaries, coupled with other coercive but non-military agencies of state power, which have significantly enhanced the appeal of hybrid conflict, not least by being able to influence audiences far beyond those who feel the immediate outcome of such efforts. Whether it be targeted assassination of key individuals using swarms of butterfly drones controlled from thousands of kilometres away, or destroying the largest ammunition dump in the world by remotely dropping chemical compounds that on combining produce molten metal at temperatures of upto 4,000 degrees Centigrade,[1] the battlespace of today defies traditional definition. It has no geographically defined frontiers, since violence can be delivered at any place, with great accuracy, from far away. Traditionally vulnerable spaces in battle or in a campaign, say, of ensuring security of one's logistic trains to the rear of the front line, have conceptually been extended in distance to many thousands of kilometres. The entire world now can be visualised to be the arena of war, where nothing is safe from attack. As for the weapons of war, these have metamorphosed to include a range of instruments covering the purely kinetic to the non-kinetic, including space, cyber and information operations, all supplemented by artificial intelligence. Furthermore, economic, political and social tools are leveraged against the adversary as part of a combined national effort designed to bend him to one's will. With options to inflict injury, death and destruction or create conditions for disorder in the target's heartland, conceptually, 'every nation's homeland is now militarily in play'.[2] Such is the reality of hybrid war.

## Effect of Tools of Hybrid Conflict on the Current Environment

It is apparent that instruments of future conflict shall progressively comprise of a greater proportion of hybrid elements. It also appears that while countries with access to high-end technology might be able to utilise the entire spectrum of instruments, others deprived of this advantage will still have options other than conventional warfighting to achieve their aims. This is borne out from a study of *Unrestricted Warfare*, the Chinese classic[3] which first entered the West's consciousness in the early 2000s. Though technology then was not what it is today, the authors make out an effective case for prosecution of national aims through hybrid means, using instruments of national power available at the time. Given the exponential advances in technology and its impact, the consequent disruption to society, and the possibility of purely military action therefore becoming just an adjunct to other means for achievement of national aims, has been discussed briefly by Lieutenant General (Lt Gen) David Barno and Nora Bensahel in an essay in 2016.[4] Against such a backdrop, it becomes necessary for India to look at its current security challenges and mull over options for fulfilment of aims through use of instruments of national policy which are not restricted to just the kinetic contact/non-contact variety possessed by the military, but encompass others, while preparing itself to counter such threats.

Pakistan's prosecution of proxy war in Jammu and Kashmir encompasses many components of hybrid warfare. Pakistan has, in fact, followed this model imaginatively; and while not having been able to wrest Kashmir by force of arms or by abetting a successful insurrection, it has definitely succeeded in keeping the pot boiling and focusing attention on the valley. Meanwhile, though Pakistan has never overtly articulated the thought, its unremitting hostility and chance remarks of its leaders have made many in India wonder whether its grand strategy vis-à-vis India goes beyond securing Kashmir for itself, since hostile actions carried out by it affect not just Kashmir but the whole country. The scope of its information warfare campaigns (disinformation for the most part, through social and print media), 'Lawfare' (by repeatedly taking up cases in international forums such as the Indus Waters Commission and the United Nations [UN]), use of fifth columnists who are regularly caught in India's hinterland and stubborn negation of any attempt to resolve differences or increase interaction with its larger neighbour lend some credence to this view, supported again by its orchestration of terrorist attacks on a regular basis inside India and involvement in criminal activities such as smuggling of fake currency. It continues to invest in setting up jihadi infrastructure and nurturing terrorists of various nationalities, while building up its conventional military strength. It has endeavoured to reduce the nuclear threshold in the sub-continent by threatening the use of tactical nuclear weapons. Given the current stalemate in relations, India will continue to face Pakistan's hybrid warfare and therefore, looking for effective counter options is a *sine qua non*.

It is necessary, at this juncture, to add a caveat. While it is not the intent in this chapter to discuss India's warfighting strategy against Pakistan, India's response (or lack thereof) to the Parliament attacks of 2001, and later the Mumbai attacks of 26 November 2008, has thrown up certain lessons, as has our response to the Uri attacks of 2016 in the form of 'surgical strikes'. These lessons must be incorporated to decide on what exactly could be the range of responses, including those of the hybrid kind, for India to activate vis-à-vis Pakistan. These could range from a full-spectrum war, short of a nuclear exchange, to coercive diplomacy using third parties for forcing a recalcitrant neighbour to come to terms. Such an exercise would have the welcome fallout of evaluating afresh the efficacy of current force structures, and shape the consequent discourse on capacity/capability building.

With respect to China, it would be realistic to accept that Sino-Indian relations are currently at a delicate stage. In a recent essay, former Foreign Secretary and Ambassador to China, Nirupama Rao, has argued that 'taking a page from our inherent capacity to adjust and be flexible, India should practice a little unemotional pragmatism.'[5] Given Chinese intransigence on the matter, the border issue will continue to fester. Rising aspirations will spur economic rivalry, which in turn will result in deployment of military force to secure essential lines of communication through which goods and services ply. Chinese bases in Djibouti and Maldives as well as acquisition of port facilities on long-term lease in Sri Lanka are some of the actions taken by that nation in the recent past, providing a strategic dimension to its Maritime Silk Route and Belt and Road project. A deep-sea port in Kyaukpyu in Myanmar is under development, and the Chinese support for Myanmar in the Rohingya crisis is seen to be aimed at freeing Rohingya land in Rakhine state for Chinese projects.[6] Chinese chequebook diplomacy, something India cannot match, is the new proven mantra for gaining influence in India's neighbourhood. China has also amassed a sizeable trade surplus with India of over 51 billion dollars, and has captured 51 percent of India's smartphone market, an extremely sensitive area for obvious reasons.[7] China also remains steadfast in its support for Pakistan, and with commencement of work on the China–Pakistan Economic Corridor (CPEC), it has shown disregard for India's sovereignty. China's recent military modernisation and moving on to the next phase of informatised warfare (including space)[8] has obvious implications for India. In its *Annual Report to Congress*, the Department of Defense of the United States (US) concludes that China:

> uses a variety of methods to acquire foreign military and dual-use technologies, including cyber theft, targeted foreign direct investment, and exploitation of the access of private Chinese nationals to such technologies. Several cases emerged in 2016 of China using its intelligence services, and employing other illicit approaches that violate U.S. laws and export controls, to obtain national security and export-restricted technologies, controlled equipment, and other materials.[9]

In addition to confronting navies of other nations in the South China Sea, the Chinese Navy now makes regular forays in the Indian Ocean and the Bay of Bengal, the new arena of potential conflict.

Considering the issues briefly outlined here and the pattern of Chinese behaviour with respect to the border, up to the Doklam incident and beyond it, realpolitik demands that India consider all options while dealing with its northern neighbour. An objective assessment of Chinese behaviour in that country's quest for realisation of the 'Chinese Dream' does not inspire confidence in the notion that good relations, or failing that, diplomacy, can always prevail, while a more extreme view might even conclude that China's hybrid war against India has commenced.

The hybrid conflicts in Eastern Europe and West Asia, covered in detail in earlier chapters of the book, also have an impact on India. While the struggle in Ukraine epitomises the hybrid form of state-on-state conflict, in West Asia, it is the non-state actors—some of them proxies for regional powers—who are the instruments. Be it the Al Qaeda or Islamic State of Iraq and Syria (ISIS), whose fighters after the fall of the caliphate are available to state and non-state sponsors of terrorism, the Houthis of Yemen (whose employment of female fighters trained by women instructors from Iran, Iraq, Syria and Lebanon is a novel feature[10]), or Pakistani Shia fighters of the Zainebiyoun Brigade in Syria,[11] these organisations continue to attract new recruits, including from South Asia. As per newspaper reports of February 2017, the National Investigation Agency (NIA) had carried out arrests pertaining to 12 cases of terrorism by ISIS,[12] while the case of entire families from Kerala moving to Afghanistan to support the Taliban is well known.¹³ Both Bangladesh and Sri Lanka have made deradicalisation campaigns their top priority in combating terrorism. For India, this 'third front' is the most insidious part of hybrid war—fighting an enemy within, whose religious or ethnic persuasions can be manipulated by state and non-state actors alike for inflicting violence through newer and deadlier instruments of terror. The attacks on innocent pedestrians in France and England using heavy vehicles offer a grim prognosis of things to come.

## The Threat

What, therefore, are the current and emerging hybrid threats which the Indian state will have to prepare for, given all that has been discussed earlier? The threat of use of kinetic military force by adversaries on both northern and western borders using contact and non-contact means (missiles, rockets and aircraft), for settling borders through warfighting, is based on live precedents. Aided by the latest C4I2SR¹⁴ means, this threat is now more potent. Construction on islands and reefs in the South China Sea continues to see steady progress, reinforces the perception of Chinese approach towards boundary disputes. At the 19th Party Congress in October 2017, timelines drawn for the Chinese military to modernise and acquire the capabilities required to become 'world class' have also been clearly articulated.¹⁵ The use of new technology for manufacturing non-kinetic weapons for war, such as anti-satellite weapons and systems for jamming civil and military communications, currently under development in China,¹⁶ coupled with cyberattack on a large Indian infrastructure company earlier this year by suspected Chinese hackers, using virtual private networks (VPNs) and proxy servers based in other countries, indicates the grim potential that such technologies possess.¹⁷ Targeted cyberattacks on various facets of the economy, such as factories, energy supply grids, including wind and solar farms, banks and railway and transport systems, described so presciently (though in a different context) by a former Indian Army Chief,¹⁸ can result in derailing governance, law and order and the economy. The effect of such disruption would be accelerated through information warfare, use of fifth columnists and other subversive means.

Attacks by terrorists, both foreign and home-grown, supported from across the border and within, shall continue to occupy the time and resources of Indian law enforcement and security agencies. As mentioned earlier, so also will the whereabouts and employment of unemployed ISIS fighters. Transborder movement of radicalised youth, some posing as refugees who can be activated later for fomenting jihad, is another threat.

In India's hinterland, the Maoist insurgency will continue to fester unless concurrent steps taken by enlightened leadership to make the indigenous people stakeholders in development and good governance are seen to bear fruit. Sabotaging India's growth through smuggling, dumping of goods, protectionism, circulating of fake currency and other unfriendly activities will require to be curbed through more ingenious methods. With respect to threats emanating from non-traditional sources, use of water as a weapon through creation of either shortages[19] or oversupply (resulting in floods), or large-scale ecological damage, is another possibility,[20] with similar problems in Pakistan occupied Kashmir (PoK).[21] And finally, India must always be alive to that most dangerous of threats: contamination of foodstuffs, medicines and natural resources by chemical and other means.

## Preparing for the Future

Faced with such challenges, what are the options for the Indian state? As always, Lord Baden-Powell's common-sense motto, 'Be Prepared', comes to mind.[22] The sheer scale of the problem outlined mandates that any preparations have to be matrix based, with strong vertical and 'cross-cultural' linkages, encompassing administrative, law enforcement and intelligence agencies at the national, state and district levels. These have to be interfaced with disaster management authorities, security forces, including Central Armed Police Forces (CAPF), science and technology organisations, infrastructure development organisations in public and private sectors (to include hygiene, water and sanitation), the health industry, customs and immigration, commerce, aviation, shipping, railways and tourism, schools, colleges and educational institutions, the media (specially social media) and many others. All elements and enterprises of the Indian state would be affected. Information and intelligence would have to be shared seamlessly so that everyone is sensitised on emerging threats, with authorities being empowered to decide when and how to act.

The Department of Homeland Security was set up in the US after the 9/11 attacks. Talk of a similar model for India has cropped up time and again,[23] with concurrent discussions on the requirement to restructure the Ministry of Home Affairs. Consequent to the report of the Kargil Review Committee, a Group of Ministers (GoM) on National Security was set up by the prime minister. In my opinion, the far-sighted recommendations of this GoM (most of which are in the open domain) remain relevant even today, as they lay out the broad framework of organisations and restructuring required for strengthening India's internal security apparatus.[24] Efforts at reorganising sensitive portions of the internal security apparatus would naturally be shielded from the public eye; however, the degree of success achieved in operationalisation of the National Intelligence Grid (NATGRID), creating an effective National Counter Terrorism Centre (NCTC),[25] coordinating efforts of intelligence and law enforcement agencies through the Multi Agency Centre (MAC) and finally, linking all these agencies on secure data links is unclear.

What has to be first decided is: whether the existing agencies just mentioned are adequate for combating the latest challenges or if a totally new organisation is required to be conceptualised and created? To arrive at a conclusion on this matter, an informed audit of the current organisations (which has already been raised) would have to be carried out, to be followed by a debate on whether these could be tweaked in some manner to cater for newer challenges. Also, keeping in mind the current organisation and system of functioning, whether or not some constitutional authority is to be vested in the head of this revamped apparatus, as happens in other countries, is another aspect that merits serious analysis.

Secondly, in a federal structure like India's where multifarious concerns require to be addressed, consensus has been the traditional norm for promulgating major reforms. This approach inevitably leads to delay and watering down of certain provisions. Considering the urgency of the situation, time-bound interaction between all stakeholders is called for, to convince them to cooperate unreservedly on this matter. Linkages between central and state agencies must be clearly delineated for efficient functioning. Channels of communication must be specified without overlapping for coordination of all types of effort from the centre down to the district level, with attendant responsibility laid down. It is likely that, finally, implementation of these suggestions might well result in concurrent restructuring at the Ministry of Home Affairs, an outcome which would be the subject of a separate study. Work on the important issue of strengthening linkages between the armed forces and other security agencies, which has commenced with the proposed raising of the tri-service defence cyber, space and special operations agencies,[26] is another focus area for speedy implementation. With the threat being seamless, this institutionalised interface will help in breaking down 'silos' between the civil and the military when tackling a threat, enabling synergy and optimisation of effort.

Enhancing security consciousness of the average citizen is the third area which needs constant monitoring. The public holds the key towards prevention of terrorist incidents or in providing information on any unusual activity. While the public awareness campaigns of the police in the wake of terrorism-related incidents have definitely met with success, additional mechanisms must be evolved given the sheer scale of hybrid warfare and the technologies involved. These could be through education in schools and colleges, well-structured media campaigns by the authorities or mock drills in public places, all carried out with deliberation and without inculcating a sense of paranoia. In any enterprise of such nature which depends on public cooperation for success, the police from the state level downwards, again, are the key players, and would first have to be appropriately trained and sensitised.

This brings us to the fourth area of focus, which involves training, equipping, modernising and increasing the numbers of our extremely overburdened police forces. Much has been written about implementation of key reforms in this regard as suggested by various commissions. As a nation, India has no choice but to progress these issues with utmost urgency. Analysis of recent terrorist attacks in France and England shows how an efficient police force can neutralise terrorists within 24 hours of an incident's occurrence, to restore calm and order. On the positive side, my interaction with senior police officers recently indicates a heightened awareness of the requirement for rejuvenation of police research and training to deal successfully with all aspects of internal security and hybrid conflict. It is hoped that this thought process will result in tangible outcomes.

This chapter had commenced with an observation about the armed forces. Enabling the military to fight hybrid war will require not just newer doctrines and capacities, but major reorganisation, as the Indian Armed Forces gradually shift away from the conventional to the newest generation of warfare. An outline of the thought process has been made with release of the first Joint Warfare Doctrine,[27] which would further mature with experience, arriving finally at an Indian version of comparative doctrines, such as the Russian doctrine[28] (which too is the outcome of an evolutionary process and is mentioned in an earlier chapter). While it is heartening to see that the military is beginning to be heard more frequently at various forums, important issues such as conflict prevention through deterrence, war aims, capacity building, consequent equipping and raising of formations and reorganisations, including tri-service integration and setting up of theatre commands, must be vigorously debated in house and options speedily arrived at. Side by side, efforts must continue to address another identified lacuna: that of arriving at a suitable model for professional education for serious military professionals.

## Conclusion

Combating hybrid warfare is a full-time task. Declarations of war have become meaningless since war can be prosecuted silently with an element of deniability, to deadly effect. The Indian people and all organs of the Indian state have to be fully alive to the threat, which has been in incubation for long and has now started manifesting itself at its time and place of choosing. The armed forces, with valued assistance from other CAPF, have traditionally been acknowledged as the sword arm of the republic. Realisation, however, has crept in that the demands of fighting hybrid war go beyond the traditional and require active participation of all segments of society, where education, state supervision, without becoming controlling, and a sense of balance to provide for the diverse needs of India's heterogeneous society are the vital pillars on which future strategies must be built.

### NOTES

1. Debashish Bose, 'A New Threat Vector', Centre for Land Warfare Studies, 28 August 2017 accessed at http://www.claws.in/1791/a-new-threat-vector-debashish-bose.html on 11 December 2017.
2. Albert Palazzo, 'Multi Domain Battle: The Echo of the Past (Part 1)', *The Strategy Bridge*, 11 October 2017 accessed at https://thestrategybridge.org/the-bridge/2017/12/6/multi-domain-battle-the-need-for-integration-with-national-strategy on 11 December 2017.
3. Qiao Liang and Wang Xiangsui, *Unrestricted Warfare*, Beijing: PLA Literature and Arts Publishing House, 1999.
4. David Barno and Nora Bensahel, 'A New Generation of Unrestricted Warfare', *War on the Rocks*, 19 April 2016 accessed at https://warontherocks.com/2016/04/a-new-generation-of-unrestricted-warfare/ on 21 Jan 2018.
5. Nirupama Rao, 'China may be an Adversary, but making it an Enemy will not Serve India', *The Wire*, 6 October 2017 accessed at https://thewire.in/184678/india-china-relations-adversary-enemy/ on 12 December 2017.
6. 'How Chinese Cash Shores up Myanmar's Rakhine State despite International Condemnation of Rohingya Crisis', *South China Morning Post/AFP*, 28 September 2017 accessed at http://www.scmp.com/news/asia/southeast-asia/article/2113259/how-chinese-cash-shores-up-myanmars-rakhine-state-despite on 13 December 2017.
7. 'China Nibbles at India: How Xiaomi, Vivo, Oppo are Driving Indian Companies Out of their Own Smartphone Market', *Financial Express*, 22 June 2017 accessed at http://www.financialexpress.com/industry/technology/china-nibbles-at-india-how-xiaomi-vivo-oppo-are-driving-indian-companies-out-of-their-own-smartphone-market/729647/ on 17 September 2017.
8. Elsa Kania and John Costello, 'China's Quest for Informatization Drives PLA Reforms', *The Diplomat*, 4 March 2017 accessed at https://thediplomat.com/2017/03/chinas-quest-for-informatization-drives-pla-reforms/ on 18 December 2017.

9. *Annual Report to Congress: Military and Security Developments Involving the People's Republic of China 2017*, Office of the Secretary of Defense, USA, p. ii.
10. 'Why are the Houthis Recruiting Female Militants', Future for Advanced Research and Studies, 16 October 2017.
11. Abdul Ghani Kakar, 'Iran-backed Militias Recruiting Pakistanis, Afghans to Fight in Syria', *Pakistan Forward*, 22 February 2017 accessed at http://pakistan.asia-ews.com/en_GB/articles/cnmi_pf/features/2017/02/22/feature-01 on 12 December 2017.
12. 'Indians Arrested so far for their Alleged Affiliation to Islamic State', *The Indian Express*, 26 February 2017.
13. Pravin Swami, 'Baby Born to Kerala Family that Joined Islamic State in Afghanistan,' *The Indian Express*, 24 August 2016.
14. C4I2SR is an acronym for Command, Control, Communications, Computers, Intelligence, Information, Surveillance and Reconnaissance in common military parlance.
15. Charlotte Gao, '3 Major Takeaways from Xi Jinping's Speech at the 19th Party Congress', *The Diplomat*, 18 October 2017 accessed at https://thediplomat.com/2017/10/3-major-takeaways-from-xi-jinpings-speech-at-the-19th-party-congress/ on 19 October 2017.
16. Harsh Vasani, 'How China is Weaponising Outer Space', *The Diplomat*, 19 January 2017 accessed at https://thediplomat.com/2017/01/how-china-is-weaponizing-outer-space/ on 21 December 2017.
17. Sachin Dave, 'Chinese Hackers Step up Attacks on Indian Shores', *The Economic Times*, 7 August 2017.
18. S. Padmanabhan, *The Writing on the Wall: India Checkmates America 2017*, New Delhi: Manas Publications, 2004, pp. 229–230.
19. Uttam Kumar Sinha, *Riverine Neighbourhood: Hydro-politics in South Asia*, New Delhi: Pentagon Press, 2016, pp. 39–40.
20. Brahma Chellaney, 'China is Waging a Water War on India', *Hindustan Times*, 21 August 2017.
21. Rashme Sehgal, 'India's Water Security Concerns over China's Dam Building Spree are Legitimate, Require Action', *Firstpost*, 18 May 2017 accessed at http://www.firstpost.com/india/indias-water-security-concerns-over-chinas-dam-building-spree-are-legitimate-require-action-3454406.html on 21 December 2017.
22. The Scout Association, *The Scout Handbook*, accessed at https://www.traditionalscouting.co.uk/documents/scout_handbook/on 05 February 2017
23. Pankaj Vohra, 'Modi may Create a Homeland Security Ministry', *Sunday Guardian*, 10 January 2016 accessed at http://www.sundayguardianlive.com/news/2558-modi-may-create-homeland-security-ministry on 17 December 2017.
24. 'Report of the Group of Ministers on National Security', pp. 41–57.
25. Sudhi Ranjan Sen, '10 Big Facts on Centre vs State over NCTC, New Anti-Terror Agency', *NDTV*, 6 May 2012 accessed at https://www.ndtv.com/india-news/10-big-facts-on-centre-vs-state-over-nctc-new-anti-terror-agency-480847 on 17 December 2017.
26. Sushant Singh, 'Coming Soon: Ministry of Defence's Cyber, Space, Special Operations Divisions', *The Indian Express*, 16 October 2017 accessed at http://indianexpress.com/article/india/coming-soon-ministry-of-defence-mods-cyber-

space-special-operations-divisions-4892404/ on 12 December 2017.
27. Headquarters Integrated Defence Staff, *Joint Doctrine: Indian Armed Forces*, New Delhi: Directorate of Doctrine, Headquarters Integrated Defence Staff, April 2017.
28. General Valery Gerasimov, Chief of the General Staff of the Russian Federation, 'The Value of Science in Anticipation: New Challenges Require Rethinking the Forms and Methods of Warfare', Rough translation by Robert Coalson [Facebook note], 21 June 2014, available at https://www.facebook.com/notes/robert-coalson/russian-military-doctrine-article-by-general-valery-gerasimov/10152184862563597/ accessed on 01 August 2017.

# 10

## Conclusion

*Neha Kohli*

---

At the end of this exhaustive volume on hybrid warfare, we find ourselves at an opportune moment to revisit the original premise behind undertaking this study. A simple question raised in a meeting—'what would be the kinds of wars that India would have to fight in the future, or the conflicts it is likely to face'—set the ball rolling. A considerable amount of discussion and debate led us to finalise two key, foundational aspects of this study: the first is a commonly agreed upon, broad definition of what constitutes hybrid warfare; and the second is the case study approach. With regard to the latter, there was a conscious decision to study not just Pakistan and its use of hybrid components vis-à-vis India and Afghanistan, and China's hybrid approach to the maritime space, but also cases elsewhere in entirely different geopolitical and geostrategic contexts with no immediate geographical connect to India. Thus, we included a discussion of Russia's application of non-linear warfare in Ukraine (Crimea), Estonia and Georgia, based on the Gerasimov Doctrine of 2013, and three case studies from West Asia that look at a decade plus of the evolution and application of hybrid warfare in the conflict-ridden countries of Syria, Iraq, Lebanon and Yemen; and also at Israel which uniquely is both a recipient and user of hybrid means against its adversaries. As far as the commonly agreed upon and applied definition of hybrid warfare goes, all authors more or less conform to Frank Hoffman's definition as applied to their individual analyses: 'a blend of the lethality of state conflict with the fanatical and protracted fervor of irregular war.'

The discussions on hybrid warfare seen in the various case studies undertaken in this volume have revealed that some elements remain common across different conflict situations, while the use of others is relevant to, and dictated by, specific events and contexts. While these have existed since the beginning of conflict in human societies, the use of hybrid elements has become more sophisticated with time, aided in large part by rapid changes in technology. Further, be it a state-on-state conflict or that between an amorphous non-state entity vis-à-vis one or more states, which could be both episodic or in continuum, the use of any hybrid component today appears magnified. This has had a far greater effect than hitherto thought of or seen, owing to the technological and information revolution of the past few decades that both drives and amplifies it. Also, the use of hybrid means has brought about a greater urbanisation of the conflict—the 9/11 attacks in New York, the 26/11 attacks in Mumbai and more recently, the ISIS-supported attacks in Europe are examples with vivid recall—with increased visibility and impact, and consequently greater humanisation with instant connect. Finally, while both state/non-state elements use hybrid means or are involved in hybrid conflict, the former often fall back on conventional means to confront such a conflict/threat that they face. Moreover, states have to hold back in many such instances as they conform to agreed-upon international norms of state behaviour, seek to avoid collateral damage, and often have domestic constraints imposed upon them, especially in democracies. In comparison, non-state entities do not have the same self-imposed limitations and therefore have greater freedom to act spectacularly. Conversely, the 'hybrid' aspect actually gives a state far greater influence and/or leeway in influencing another state or limiting its actions vis-à-vis a response.

The book begins with two incisive conceptual chapters that introduce the reader to what constitutes war and conflict today and, within that context, how hybrid warfare is defined. In a globalised, interconnected world that is hurtling down the information highway, the roots of conflict lie in the weathering down of the nation-state system, in glaring (and rising) socio-economic inequalities, in the increasing defining of the self and community in terms of ethnic and religious identities, in crises of governance, in scarcity of food, water and natural resources, and in climate change and migration, to name but a few. The majority of these are no longer limited by national or territorial boundaries but transcend them, and the players include both defined states and protean non-state actors. Thus, the war in Syria and the emergence of the Islamic State of Iraq and Syria (ISIS) in West Asia reverberates in Europe.

Even if there is no clear, apparent conflict scenario, there are simmering undercurrents that are both episodic or in continuum, as the various analyses show. Hybrid warfare in itself is not unique to our times; rather, taken as a combination of both kinetic and non-kinetic tools, it has existed in all documented forms of conflict throughout millennia. Studies focusing on indigenous, non-Occidental/Western political theory and military science, such as that on Kautilya's *Arthashastra* and Sun Tzu's *The Art of War*, have indicated clearly that hybridity, gray zones and non-linear approaches existed throughout the history of Oriental political and military science. The distinction, if any, comes from the technological age and the socio-political context in which the conflict occurs—for example, the visibility and impact of any event that can be characterised as an outcome of hybrid warfare is today amplified by virtue of our living through a near-constant digital and information revolution. Depending on the perspective and situation, it has been termed as hybrid, gray zone, unrestricted warfare or non-linear conflict, but its essential character remains definitive. Thus, 'hybrid' as a term can be applied to both a declared or undeclared conflict situation where two or more of the elements of power are utilised or applied consciously, concurrently and systematically to achieve the desired political and strategic goals.

The information, or digital, revolution has enhanced the use of hybrid means and made it easier for both states and non-state actors to undertake hybrid warfare against an adversary. The levels of sophistication and the intensity, however, vary and its effectiveness or success is open to debate, or perhaps discernible in the longer term. Pakistan's hybrid warfare against its neighbours, India and Afghanistan, when seen from a long-term perspective, distinctly shows how the former has always used hybrid means to achieve its defined strategic goals, even though the goals as well as the means have evolved and been refined over the decades. In the case of India, Pakistan has used hybrid means against a conventionally (militarily) superior neighbour. Using these means enables Pakistan to undercut this conventional superiority in its favour. This is especially the case with the nuclear overhang, which makes an out-and-out conflict between the two South Asian neighbours rather unlikely. In the case of Afghanistan, Pakistan uses hybrid elements against a considerably inferior opponent, and these are geared towards increasing and maintaining its influence in that country, aided considerably by geography and geopolitics.

Our examination of Russia's use of hybrid tools in three instances indicates that it has been extremely successful in applying the concepts of hybrid war, as expounded by Gerasimov, in furthering its strategic and political aims. The Russians have been adept in changing their means and methods as required to achieve their aims, especially by exploiting non-military/non-kinetic means. Their success thus resulted in the minimal use of military means while achieving the stated strategic goal. This enabled escalation control by restricting the conflict to Russia's advantage. The Russians are very clear about the desired end state and can withstand extreme diplomatic pressure while striving to achieve it. The management of information and control over dissemination was aptly demonstrated in Crimea by denying their involvement and disowning the 'little green men' while the crisis unfolded, and accepting that they were their armed forces personnel later. Moreover, a readiness to use proxies as required and follow up with the conventional force of arms at the appropriate time epitomises Russian readiness to adapt to the evolving battlefield. At the same time, applying an economy of effort by combining kinetic and non-kinetic means in order to achieve their goals underlines what hybrid warfare stands for.

The first of the three West Asia-focused cases studies is on Iraq and Syria and the ISIS. The American involvement in the region is long-standing, and especially so in the aftermath of the 2003 United States (US)-led invasion of Iraq that caused a radical shift in regional geopolitics and paved the way for the emergence of entities like the ISIS. The US' involvement in the region has only deepened and finds mention in their strategic outlook towards West Asia. Our focus on this highlights the changes brought about in the US military thinking stemming from the need, as expounded by Jim Mattis, to focus on non-conventional means while retaining normative military competence. The present Capstone Concept evolved from this thought process, focused on a joint operating environment in which US forces would operate to protect American interests with constrained resources, while facing adversaries with increasing hybrid warfare capabilities. This thought process was further catalysed and analysed with the advent of ISIS and Russia's action in Crimea. It is in this context that the application of hybrid warfare tactics by the US in Iraq and Syria has been analysed. Furthermore, the rise of the ISIS and its dexterity in adapting to the changing nature of the battlefield and methods used by its adversaries highlights its ability to exploit the grid that relates variables defining hybrid threats to their impact on conflict and response. The six lessons derived from the American experience can be scrutinised in the Indian context and aspects meriting attention addressed accordingly.

The second West Asian case study, focusing on the conflicts in Lebanon (2006) and in Yemen (2014 onwards), reiterates that the geopolitical contestation for regional superiority between Iran and Saudi Arabia, and

Israel versus its Arab neighbourhood and Iran, manifests itself in hybrid terms. Thus, we see the prevalence of proxies with external state funding and support; engagement of proxies with conventional militaries—as was the case with the Hezbollah and Israel in Lebanon and can be seen in Yemen with the Saudi-led Sunni coalition vis-à-vis the Houthi rebels, terrorism; use of economic instruments; and so on. State sponsorship of proxies in this case is primarily the confirmed Iranian backing for the Hezbollah and the alleged backing to the Yemeni Houthi rebels. State sponsorship, declared or otherwise, brought these proxies an infusion of money and military technology; and in a region that has seen a rapid degeneration of political authority it is also notable that, in conflict situations, these proxies have in the two cases bested conventionally and technologically superior military forces. This was especially the case in Lebanon where Hezbollah, despite having a larger casualty rate and finding its military strength eroded, actually won the perception battle vis-à-vis Israel. In Yemen, we see an added layer of conflict complication in the form of the Al Qaeda in the Arabian Peninsula (AQAP) factor, which the US is targeting simultaneously—for example, the over 80 airstrikes that the latter has conducted against the AQAP in Yemen in 2017. In this murky scenario, the hybridity of conflict is indicated in an obscuring of roles played by states and their proxies. The latter are bound less by the rules and norms that bind states, and perhaps this nebulous character allows greater space for co-option of hybrid means by all players.

Israel, as mentioned earlier, has been the target of hybrid warfare as well as an exponent, having utilised similar tactics against its opponents. Three instances of conflict—direct and ranging from low to high intensity, in the 2006 war with Hezbollah; the three Gaza conflicts since 2005; and an ongoing non-contact conflict with Iran—that are discussed show that it has tackled these with varying degrees of success. In the case of Hezbollah and Hamas, the Israeli experience underscores the effectiveness of its conventional capability to tackle the significant wherewithal such hybrid groups bring into the conflict situation. What emerges from this study is that the 2006 war is 'a cautionary tale of the continuing relevance of effective training and preparations that need to be made as regards the conventional capabilities of a nation-state, against enemy capabilities that range from the kinetic to the non-kinetic spectrum.' Despite its inventiveness in successfully executing tactics specific to a particular security situation, Israel was unable to notch up a political victory in Lebanon, which in turn forced it to 're-evaluate its use of force philosophy and practice'. The lesson that its experience can provide to other nations grappling with a hybrid threat is the effective and profitable use of technology, stitched into the more specific strategic objective that is the desired end result.

The maritime environment is the new frontier in the strategic ambitions of aspiring powers, and China's assertiveness in this space—especially in the East China Sea and the South China Sea—has all the hallmarks of a hybrid or gray zone conflict. Buoyed by its economic growth for over two decades in double-digit figures, China is aiming to expand its claim on power and resources 'through a variety of measures, political, diplomatic and military, including coercion of potential opponents through limited escalation and delayed resolution of issues which it could not settle in its favour as yet.' Its strategy has been to take small, yet incremental, steps to strengthen and embellish its jurisdiction claims in the near maritime domain. These, however, stop short of or are well below what could be considered as an escalation into an open conflict. Circuitous as well as consistent, this approach appears grounded in classic Chinese thought, revealing an unobtrusive, long-term approach to manipulate the regional strategic configuration in its favour. If we look at its activities in the Indian Ocean Region as well, China's moves in terms of resource acquisition, coupled with its approach towards conflict management in the East and South China Seas, are indicative that it can easily tailor its methods to specific situations and objectives even beyond its maritime periphery.

The purpose behind analysing these specific case studies was to extract lessons for approaching the wars of the future from an Indian perspective. Chapter 9 thus highlights the need for the armed forces to continuously evaluate the changing dynamics of the operating environment, adapt accordingly, and consistently formulate appropriate strategies to address the same. In the case of hybrid warfare, this is an imperative as hybrid elements are increasing at a rapid rate and are more evident in modern conflicts. In the context of Pakistan's proxy war against India, the hybrid means adopted address its conventional, technological and economic inferiority. On the other hand, despite being conventionally superior with far greater economic heft and larger global ambitions, China's India strategy showcases a different methodology of the use of hybrid elements. What emerges thus is that conventional superiority or inferiority has no bearing on a state's use of hybrid means against an adversary.

The differing aspects of hybrid warfare faced by India highlight some fundamental issues that must be (re)considered by practitioners while devising a response. The explorations in this volume have indicated clearly the kinds of wars/conflicts that India is likely to face in the future.

Addressing hybrid forms of warfare or conflict would, at the foundational level, necessitate a shift in the doctrinal or conceptual aspects of warfighting. Therefore, there is an urgent requirement to go beyond the vertical, silo-based thinking that characterises our approach to contemporary wars and conflict and *reconceptualise the theories of warfare*. Flowing from that would entail addressing the associated aspects of strategies and capability development of the armed forces. The current approach to transformation emphasises more on new technologies and modern science, involves focus on cyber, artificial intelligence and the like. However, it is equally important to study and understand the political and socio-economic aspects of conflict, not just in how they determine the origin of the conflict but also in how we approach its mitigation. This acquires criticality as conflicts can no longer be limited in terms of time and space, especially in the information age.

Thus this volume, in particular Chapter 9, points to future areas of research that scholars as well as practitioners of military and defence studies should focus on. Recalibrating both individual as well as institutional thinking would begin with conceptual clarity, wherein the static diffusion of warrior/civilian/scholar etc., and a silo-based thinking must be done away with if we are to realise and address the looming threat that hybrid warfare poses. Furthermore, the fundamental approach to warfighting therefore would increasingly have to be geared towards strategies for long-term *conflict management*, rather than merely limited to fighting wars in the hope that their culmination would mean an end to conflict as well. We hope that this volume on hybrid warfare has fulfilled its goal of presenting a uniquely Indian perspective and sets the tone for further exploration on the subject.

# Index

19th Party Congress, 178
2001 terrorist attacks, 2, 11, 19, 186

AAA batteries, 112
Abu Ali Mustafa Brigades, 126
Actual Ground Position Line (AGPL), 18
Administered peacefully and stably, 150
Admiral Harris, 141
Afghan Taliban, 50
Afghanistan, 20, 37, 43, 48, 50-52, 55-58, 74, 187
Afghan-Pakistan border, 50
Ahmed al-Jabari, 126
Ahmed, Ismail Ould Cheikh, 114
Air Defence Identification Zone (ADIZ), 151, 157
Air Power, 111
Airborne Early Warning and Control (AEW&C) system, 105
Air-to-Air Refuelling, 105
Akhnoor, 41-42
Al Aqsa Martyrs Brigades, 126
*Al Manar*, 103, 130
Al Qaeda in the Arabian Peninsula (AQAP), 78, 105, 107, 109-10, 114-15, 189
Al Qaeda, 50, 52, 93, 110, 126, 134, 177
Al Qaeda-linked networks, 86
al-Assad regime, 88
Al-Baydah, 107
Al-Islah, 105-06, 115
Allon, Yigal, 124
American Revolution, 122
Amir-ul Momineen, 49
*Annual Report to Congress*, 177
Ansar Allah, 104
Anti-aircraft Artillery (AAA) batteries, 109
Anti-ship missiles (ASMs), 106

Anti-tank guided missile (ATGM) systems, 102, 106, 108-09, 112
Antonio Guterres, 133
Arab League, 104
Arab-Israel rivalry/wars, 13, 99
Arab-language Propaganda, 82
Armed Intruders, 57
Armoured Infantry Fighting Vehicle (AIFV), 109
Armoured Personnel Carrier (APC), 109, 133
Arms Control, Verification and Compliance (AVC), 9
Army of Islam, 126
Asia Maritime Transparency Initiative, 155
Asia-Pacific Economic Cooperation (APEC), 3
Association of Southeast Asian Nations (ASEAN), 3, 155, 158
Attack on India's Parliament, 46
Automated Teller Machines (ATMs), 67
Ayyash, Yahya, 126

Bahrain, 105
Bangladesh, 178
Bashar al-Assad, 90
Battle of Troy, 26
Battlespace, 116
Bay of Bengal, 177
Be Prepared, 179
Belt and Road Initiative, 161, 176
*bheda* (dissension/information operations), 33
Bhutan, 159
Black Sea, 64, 68
Blended modalities, 79
Boesche, Roger, 32
Bond, 30-31

# Index

Border Action Teams (BATs), 53
Brigadier (later Major General) Akbar Khan, 39
British Army, 7
British Defence Doctrine, 6, 10

C4I2SR, 178
Cabbage Strategy, 150, 154, 157
Capstone Concept, 75
Catherine II, 62
*CCTV*, 147, 160
CENTCOM, 91
Center for Strategic Counterterrorism Communications (CSCC), 82
Central Armed Police Forces (CAPF), 179
Central Intelligence Agency (CIA), 52, 91
Central Military Commission (CMC), 146-47
Charge of the Light Brigade, 64
Che Guevara, 7
Chellaney, Brahma, 158
Chemical, Biological, Radiological and Nuclear (CRBN) Warfare, 80
Chi Haotian, 146
China Marine Surveillance force, 150
China National Offshore Oil Corporation, 154
China, 17, 29, 88, 142, 149, 151, 153-54, 156, 159, 176
China's aggressive maritime irredentism, 141
China's Defence White Paper, 2006, 147
China's Gray Zone Strategy,
   Actions, 156
   Approach,
      in Maritime Disputes, 148-51, 152-56
   Assessing Effectiveness of, 156-58
   Concepts, 142-48
   Future Contours, 160-61
   Incidents, 157
China's Hybrid War against India, 177
China's Hybrid Warfare Approach, 151
China's Strategic Thinking, 143
China-Pakistan Economic Corridor (CPEC), 177
Chinese Classic, 175
Chinese Dream, 177
Chinese government, 146
Chinese Navy, 177
Chinese Research Ship, 150
Chinese Sovereignty, 156
Civil Disobedience, 19
Civil War, 6
Clandestine/Silent War, 32
Clash of Civilisations, 2
Clausewitz, 18, 21
Cold Test, 128
Cold War, 1, 2, 19, 48
Col 'Imam' Sultan Amir, 49
Col Dennis Drew, 15
Col Gabriel Bonnet, 6
Col Qiao Liang, 31
Col Wang Xiangsui, 31
Combination Warfare, 146
Commander of the Faithful, 49
Commercial off-the-shelf (COTS) Technology, 100
Communication Technology, 100
Communications Intelligence (COMINT), 103
Communications, 100
Communist Party of China (CPC) Central Committee, 147
Communist Revolution in China, 122
Compound Warfare, 16, 122
Computer Emergency Response Team (CERT), 67
Conflict, 5
Conventional Force, 57
Conventional, 29
Coordination, 101
Counter-insurgency, 8, 19
Creveld, Martin van, 21
Crimea, 63-66, 71-72, 75, 185, 188
Crimean Parliament, 65
Crimean Peninsula, 64
Criminality, 79
Crummy Little Wars, 3
Cuarteron Reef, 154
Cyber, 57, 100
   Warriors, 45
Cyberattacks, 66, 68-69
Cybersecurity, 4

*dana* (economic gratification), 33
*danda* (use of force), 33
David Lai, 143
Defence of Japan 2016, 113
Democratic Front for the Liberation of Palestine, 126
Diaoyu Islands, 149-51
Diplomacy, 55, 57
Disintegration strategy, 155

Disintegration warfare, 148
Disregard for International Law, 79, 82
Distributed Denial-of-service (DDOS) attacks, 67, 69
Dixit, J.N., 42
Djibouti, 176
Doklam Plateau, 158
Doklam standoff, 158
Drug trafficking, 76
Durand Line, 53, 54

East China Sea, 141-42, 150-51, 156-57, 190
  Gray Zone Dynamics, 149-51
East Jerusalem, 124
Economic Elements, 57
Eduard Shevardnadze, 68
Effects-based operations, 125
Egypt, 105, 125
Ehud Barak, 134
Electromagnetic Spectrum, 100
Electronic Intelligence (ELINT), 103, 105
Erstwhile Soviet Union, 68, 70-71
Estonia, 63, 66-67
Ethnic Tensions, 4
Europe, 79, 88
European Union (EU), 3, 66, 75, 104
Extortion and Kidnapping, 76
Eyeball-to-eyeball, 18

Fair, Christine, 38
Fake Indian Currency Notes (FICN), 45, 55
Falah-e-Insaniat Foundation (FIF), 55
Far abroad ring, 79
Fatah Al Islam, 126
Fatah-led Palestinian Authority, 133
Field Marshal von Paulus, 173
Fiery Cross Reef, 154-55
First Chinese-Communist War, 13
First Gulf War, 146
First World War, 13
Flames of War, 82
*Fleet Marine Forces Manual* 1, 13
Flexible Structures, 79-80
Force-on-force, 100
Fourth Generation Warfare, 127
Free Syrian Army (FSA), 85
Freedom fight, 43
Full-spectrum conflict, 33, 34
Funding terrorism, 45
Fusion, 79

Gates, Robert M., 1
Gaven Reef, 154
Gaza Conflicts, 123, 189
  post-2005 Disengagement, 131-34
Gaza Strip, 123, 127, 130, 133
Gen Alfred M. Gray, 13
Gen Gerasimov, 31
Gen Halutz, 131
Gen John R. Galvin, 15
Gen Joseph Votel, 91
Gen Raad Hamdani, 77
Georgia, 67-70, 185
Georgian
  Army, 68
  Forces, 70
Gerasimov, Valery, 26
  Doctrine, 63, 185
Glenn, Dr. Russell W., 17
Global Media Network, 147
Global War against Terror, 50
*Go*, 142, 144
Golan Heights, 124
Gradualist, 25
Gray Zone, 8-9, 19, 25, 27, 29, 31, 35, 141, 151
  Conflicts, 29, 30
  Stratagem: Chinese Military Classics Literature, 143-45
  Strategy, 145
Great War in La Plata, 13
Group of Ministers (GoM) on National Security, 179
Guantanamo Bay, 29
*Guerrilla* Warfare, 12, 19, 20, 112
Gulf Cooperation Council (GCC), 104, 114, 115
Gulf War, 20, 125

Haganah, 123
Hal Brands, 141
Hamas, 126, 189
  Bomb Maker, 126
  Rocket Inventory, 132
Haqqani Network, 51, 52
Harkat-ul-Ansar, 49
Harkat-ul-Mujahideen, 49
Haykel, Bernard, 104
Hezbollah, 88, 101-4, 107, 110-16, 126-27, 129-30, 132, 135-36, 189
  Defensive Bunker Systems, 102
  Finance Sources, 104
  Main Offensive Inventory Boasted, 102

Military Wing, 111
Robust Media Propaganda, 130
Tactics and Practices, 16
High Intensity Conflict (HIC), 13-14, 123, 125, 131
Hizbul Mujahideen (HM), 43
Hoffman, Frank, 16, 29-30, 74, 76, 79, 101, 142, 185
Houthi-Hadi conflict, 109
Houthis, 105-06, 108, 110-11, 113, 115-16
Huntington, Samuel, 2
Hussein Badreddin al-Houthi, 104
Hussein, Saddam, 77
Hybrid Warfare, 15-16, 19, 25, 29-31, 33-34, 38, 46, 53, 57, 69, 74-75, 77, 79, 101, 115, 174, 182, 187, 189
  Actors and Interests in Syria's, 85-92
  Advantages for Russia, 70-71
  Challenges, 84
  Conflict, 15-16, 31, 42, 177
    Components, 33-34
  Countering, 75
  Employment of, 52, 56
  Force, 75
  Forms of Warfare, 191
  in Afghanistan, 56
  in *Arthashastra*, 32-33, 35
  Means, 175
  Syrian Civil War, 85
  Theory, 35
  Threat, 15-16, 79, 84, 93, 178
Hybridity, 35
Hyper-strategic Weapons, 146

Impossible Political Goals, 132
Improvised Explosive Devices (IEDs), 80, 107-09
India, 159, 178
India's Military, 174
*Indian Army Doctrine*, 18-19
Indian Army, 14, 18
Indian Ocean, 160, 177
Indian Peninsula, Communal Riots, 13
Indochina Wars, 122
Indo-Pak War, 13
  1947-48, 39-40, 53-54
  1965, 40-42, 53-54
Infantry Combat Vehicles (ICVs), 109
Information Age, 28-29, 34, 191
  Conflicts, 34
  Tools, 44

Information and Psychological Operations, 100
Information Operations, 77
Information Warfare, 79, 82
  Campaigns, 175
*INS Hanit*, 136
Insurgency, 7-8, 19
Intelligence Department of General Staff Headquarters, 148
Interior Ring, 79
Internal Security, 19
International Atomic Energy Agency (IAEA), 134
International Security Advisory Board (ISAB), 27, 30
Internet, 66, 69
Inter-Services Intelligence (ISI), 18, 45, 48, 49, 51, 54
Iran, 86, 88-89, 104-05, 114-15, 123, 134, 177, 188
Iran's Testing of Rockets, 136
Iranian Nuclear Challenge, 123, 134-35
Iran-Saudi Conflict, 99
Iraq, 20, 74, 76-79, 83, 177, 185, 188
Iraq's State Board for Antiquities and Heritage, 83
Iraqi Army 'blitzkrieg', 78
Iron Dome AMD system, 133, 136-37
Irregular Warfare, 6-7, 29, 77
Islamic State (IS), 87
Islamic State of Iraq and Syria (ISIS), 10, 75, 76-79, 80-84, 87, 91, 93, 115, 177, 186, 188
  Militia/Forces, 16, 19
  Operations, 81
Israel, 110-11, 123-24, 131, 137, 189
  Air Force, 104, 112
  Government, 126
  Nuclear Weapons Programme, 128
  Offense Forces, 124
  Security Agency, 126
Israel Defense Forces (IDF), 102-03, 111, 123-24, 126-27, 132
  Communications, 103
  Humvees, 129
Israel's Raid on Osiraq, 125
Israel-Hezbollah War, 122-23, 129-31
Izz al-Din al-Qassam Brigades, 126

Jamaat-ud-Dawa (JuD), 55
James 'Jim' Mattis, 74
Jammu and Kashmir (J&K), 7, 18, 39-40, 46, 56

Jammu and Kashmir Liberation Front (JKLF), 43
Japan, 88, 151
Japanese Coast Guard (JCG), 150
Japanese Exclusive Economic Zone, 149
Jerusalem Brigades, 126
Jihadis, 18
Johnson South Reef, 154
Joint Comprehensive Plan of Action (JCPOA), 136
Joint Operating Environment 2008, 75
Joint Warfare Doctrine, 181
Jonsson, Oscar, 32
Jordan, 105, 125
Justified Act of Self-defence, 151

Kargil conflict, 44, 46, 50, 53
*Kargil Review Committee*, 179
Kashmir Conflict, 43
Kashmir, 41-47, 55-56, 57
Kashmiri Terrorists, 19
Kaufman, Stuart, 76
Kautilya's *Arthashastra*, 187
Kennedy, John F., 8
Kenya, 49
Khalistan, 18
Khan, Akbar, 39
Khan, Ayub, 41
Khan, Liaquat Ali, 39
Khan, Riaz Mohammad, 50
Khan, Sardar Ibrahim, 39
Kinetic Components, 33
Kinetic, 34, 72, 81, 116, 187
Kippur War, 124
Kissinger, Henry, 142
Kitson, Frank, 7
Kivimaki, Timo, 115
Krauthammer, Charles, 2
Kurdish Peshmergas, 84
Kurdistan Regional Government, 83
*Kutayuddha*/Concealed Warfare, 32
Kuwait, 105

Lashkar-e-Taiba (LeT), 55
Lawfare, 71, 147, 175
Lebanese Armed Forces, 113
Lebanon War, 123, 125
Lebanon, 99, 113-14, 123, 177, 185, 188-89
Legal Warfare, 147
Lenin, 62
Liang, Qiao, 18

Liberation of Crimea, 71
Lidell Hart, 29
Line of Control (LoC), 41, 43, 47, 53
Little Green Men, 65, 72, 188
Little Satan, 134
Logistical Network, 108
Lord Baden-Powell, 179
Low Intensity Conflict (LIC), 13-15, 20-21, 123, 125, 131
Low Intensity, 21
  Limited War, 18
Lt Col John Fulton, 15
Lt Gen David Barno, 175
Lt Gen Harbakhsh Singh,
  *War Despatches*, 41
Lt Gen Mahmud Ahmed, 50
Lt Gen Michael Vane, 93

Majlis an-Nuwwab, 113
Major Amin, 40
Major Khurshid Anwar (Retd), 39
Major Zaman Kiani, 39
Maldives, 176
Manfaz al-Khadra, 108
Man-portable Air Defence Systems (MANPADS), 109, 112
*Mantrayuddha*/War by Counsel, 32
Mao Tse-tung, 7
Mao Zedong, 148
Maoist Insurgency, 179
Marighela, Carlos, 7
Maritime
  Environment, 190
  Gray Zone Conflicts, 142
  Security Agencies, 160
  Silk Route, 161, 176
Mattis, 75
Mavi Marmara incident, 133
Max Boot, 12
Mazar-e-Sharif, 48-49
Mazzar, Michael, 142
Media Warfare, 147, 155
Medium Intensity Conflict (MIC), 13
Merkava main battle tank, 131
Merriam-Webster, 26
Mian Iftikhar-ud-Din, 39
Middle East, 79, 84
Militant, 7
Military Employment, 113
Military-Industrial Kurier, 63
Mischief Reef, 154-55

Miyako Strait, 150
Modern Conflict, 4-6
Moe Muqaddas, 41
Mohammad, Ghulam, 39
Morocco, 105
Mowing the grass, 135
*Mujahideen*, 43, 48
Multi Agency Centre (MAC), 180
Mumbai attack, 46, 176, 186
Munich Security Conference, 75
Muzaffarabad-Srinagar axis, 39
Myanmar, 161, 176

Nansha/Spratly Islands, 155
Napoleonic Wars, 122
Nation's Armed Forces, 5
National Counter Terrorism Centre (NCTC), 180
National Intelligence Grid (NATGRID), 180
National Investigation Agency (NIA), 177
National Security Agency (NSA), 135
Netanyahu, Benjamin, 134
Neumann, Peter R., 10-11
New Terrorism, 10-11
Night Vision Goggles (NVG), 103
Night Wolves, 66, 71
No War-No Peace (NWNP), 18
Non-contact Warfare, 4
Non-governmental Organisations (NGOs), 112
Non-linear Warfare, 17, 31-34, 63, 67, 72, 81, 116, 187
Non-military War Operations, 145
Non-state Actor, 3, 101, 177
Non-state Armed Group (NSAG), 122, 129
Non-state, 137
    Threats, 126-27
North Africa, 79
North Atlantic Treaty Organization (NATO), 1, 66-67, 74-75, 93
North West Frontier Province (NWFP), 48
Nuclear Threat, 57
Nuclear Weapons, 55

Obama administration, 83
Obama Doctrine 2014, 83
Omar, Mullah, 49
One Vision-One Identity-One Community, 155
Operation Al-Hazm Storm, 105
Operation Cast Lead, 131-32

Operation Change of Direction, 101
Operation Defensive Shield, 127
Operation Gibraltar, 41
Operation Pillar of Defence, 131
Operation Protective Edge, 131
Operation Summer Rains, 131, 133
Operations Other Than War (OOTW), 15
Organisation of Islamic Cooperation (OIC), 44
Organised Criminal Activity, 79
Oslo Accords, 133

Pakistan, 18, 37, 41, 43-44, 50-52, 54, 105, 161, 176, 188
    Components of Hybrid Warfare Employed by, 57
Pakistan's
    Army, 18, 39, 51-52
    Supported Sikh Militancy, 18
    Hybrid War against India, 38-47
    Hybrid War in Afghanistan, 47-52
    Hybrid War, 52-56, 58
    Madrasas, 48-49
    Shia Fighters, 177
    Taliban, 50
Pakistan occupied Kashmir (PoK), 39, 179
Palestine Liberation Organization (PLO), 125, 127
Palestinian Groups, 135
Palestinian Islamic Jihad, 126, 129, 136
Palestinian Terror Groups, 136
Paracel Islands, 154
Parliament Attacks, 176
Peacekeeping, 70
Peloponnesian Wars, 26
People's Army, 126
People's Liberation Army (PLA), 146, 148, 158
    Nanjing International Relation College, 148
    Political Work Regulations, 147
Peshawar, 48
Philippines, 154
Pillsbury, Michael, 145
PLA Navy (PLAN), 150-51, 160
Popular Front for the Liberation of Palestine (PFLP), 126
Popular Resistance Committees (PRC), 126
post-Cold War, 3
post-ISIS, 84
post-Kargil Environment, 46

Post-national State, 2
post-war Syria, 88
Powell Doctrine, 20
*Prakasayuddha*/Open Warfare, 32
Pre-emptive war, 124
President Putin, 65
President Trump, 91
Propensity of Things, 144
Prophet Mohammad, 41
Protective Edge, 132, 133
Protracted Social Conflict, 126
Proxy War, 18
Psychological Warfare, 81, 147, 155
Punjab Police, 18
Purchasing Power Parity (PPP), 62

Qatar, 86, 105, 114-15
Qazi Humayun, 49

Radical extremism, 4
Raisina Dialogue 2018, 141
Rantissi, Abdul Aziz, 126
Rao, Nirupama, 176
Rashid, Ahmed, 48
Reconnaissance, 105
Red Army, 67
Red Sea, 108
Religious Fundamentalism, 4
Responsible Stakeholder, 151
Revolutionary Warfare, 6-7, 19
Rose Revolution, 68
Royal Bhutanese Army, 158
Rules of War, 63
Russia, 62, 66-67, 69, 71, 87-89
Russia's Use of Information Warfare, 71
Russia-Iran-Turkey trio, 87
Russian
  Air Force, 68
  Armed Forces, 70
  Doctrine, 181
  Flag, 65
  Hackers, 69
  Imagination and Self-awareness, 64
  Navy Black Sea Fleet, 64
  Night Wolves, 65
  Non-linear warfare, 66
Russo-Georgian War, 68
Russo-Turkish War, 64

Saakashvili, Mikheil, 68

Salah al-Din Brigades, 126
Salami slicing, 157
*sama* (conciliation or diplomacy), 33
Saudi Arabia, 86, 105, 108, 115, 188
Saudi Forces, 105
Saudi-Iran Rivalry, 99
Scarborough Shoal, 154
Schroefl, Josef, 76
Second Gulf War, 146
Second Lebanon War, 101-04, 113
Second World War, 14, 20, 64, 67
Seely, Robert, 32
Senkaku Islands, 149-51
Sequel to the Bolshevik Revolution, 13
Shadow globalization, 76
Shanghai Cooperation Organisation (SCO), 3
Sharif, Nawaz, 50
Sheik Hamid al-Ahmar, 106
*Shi*, 144-45
Shia Iran, 89
Shiite Crescent, 115
Shinrikyo, Aum, 10
Siachen Glacier, 18, 44
Signals Intelligence (SIGINT), 103
Simferopol, 65
Simultaneity, 79
Sinai Campaign, 123-24
*Six Secret Teachings*, 144, 146
Six-Day War, 128
Small Arms Low Calibre Weapons (SALW), 100, 109
Small Stick Diplomacy, 154
Small War, 12, 19
  Tactics, 135
*Small Wars Journal*, 12
Smith, General Rupert,
  *The Utility of Force*, 26
Socio-economic Disparities, 4
South Asian Association for Regional Cooperation (SAARC), 3
South China Sea, 141-42, 148, 150, 153, 155-57, 177, 190
  Gray Zone Dynamics in, 151, 153-56
South Korea, 151
Southern Asia, 3, 4
Soviet Black Sea Fleet, 64
Soviet Union, Break-up, 2
Spanish Civil War, 13
Special Cells, 45
Special Operations Command, 9

Special Operations Forces (SOF), 27
Sri Lanka, 161, 176, 178
State versus state military conflicts, 100
State-centric Threats, 123-25, 137
State-on-state Warfare, 174
Strategy of Gray Zone Conflict, 145-48
Subi Reef, 154-55
Subversion, 19, 57
Sudan, 105
Suez War, 127
Sun Tzu, 33, 35, 143, 145, 147
  *Art of War*, 142, 144, 187
Sunni Axis, 115
Sunni Turkey in Syria, 89
Supra-domain Combinations, 146
Supra-means Combinations, 146
Supra-national Combinations, 146
Supra-tier Combinations, 146
Supreme Council (Parliament) of Crimea, 66
Supreme Military Council (SMC), 90
Surface-to-air Guided Weapons (SAGWs), 106, 109
Surface-to-surface Missiles (SSMs), 106, 108-09
Surgical Strikes, 176
Surkov, Vladislav, 31
Swift and Definite stop, 134
Swords, 25
Syria, 62, 76-77, 79, 89, 104, 114, 177, 185, 188
Syrian
  Civil War, 90, 92
  Democratic Forces, 86
  Syrian Government, 86
  National Council, 85

Taipei-based newspaper, 155
Taiping Rebellion, 13
Taiwan, 149, 151
Taliban, 49-52, 54
Taliban-controlled Drug Network, 52
Tanzania, 49
Tashkent, 48
Techno-centric Warfare, 125
Technology, 108
Territorial disputes, 4
Terrorism, 10, 19, 57, 79, 81
Terrorist, 29
*The Science of Military Strategy*, 145
*The Seven Military Classics of Ancient China*, 144
Third World War, 2
Third World, 12

Threats, 117
Tillerson, Rex, 84
Tora Bora, 52
Total War, 18
Trans-Pacific Partnership (TPP), 3
Trojan Horse, 26
Turkey, 68, 86, 88-89

Ukraine, 63, 64, 65, 71, 177, 185
Ukrainian
  Government, 65
  Naval Forces, 64
  Revolution, 64, 65
UN General Assembly, 134
UN Secretary-General's High-level Panel on Threats, Challenges and Change, 10
UN Security Council, 70
Union of Soviet Socialist Republics (USSR), 64
Unipolar Moment, 2
United Arab Emirates (UAE), 105, 107
United Nations (UN), 2, 49, 81, 101, 151, 175
  Charter, 111
  General Assembly, 44
  Scrutiny, 70
United Nations Convention on the Law of the Sea (UNCLOS), 160
United Nations Interim Force in Lebanon (UNIFIL), 101
United Nations Security Council (UNSC), 55
  Resolution 1701, 113
United States (US), 2, 20, 27, 29-30, 43, 47-51, 55-56, 62, 76-77, 79, 85, 105, 109-10, 114, 124, 135, 144, 149, 177, 188
  Army War College, 15
  Civil War, 13
  Congress, 84
  Congressional Research Service, 90
  Department of Defense, 14
  Experiences in Iraq, 75
  Marine Corps Manual on Small Wars, 12
  Military Thinkers, 78
  on Hybrid Wars, 93
  Special Operation Command, 114
Unmanned Aerial Vehicles (UAVs), 84, 102-03, 106, 108-09, 122, 127
Unrestricted Warfare, 17, 34, 145, 146, 175
Uotsuri/Diaoyu Dao, 149
Uri Attacks, 176
*USS Mason*, 108
Uzbekistan, 48

Vehicle-Borne IEDs (VBIEDs), 80
Videos, 45
Vietnam, 154
Violence, 6
Violent, 13
Virtual Private Networks (VPNs), 178
Vo Nguyen Giap, 7

Walter Laqueur, 10
War, 25-26
War Economies, 76
War of Attrition, 123-24
War of Independence, 124
Warfare, 113
Warfare (4GW), 8, 19
Warfare Beyond Bounds, 17
Wars, 27
    of Conscience, 3
    of Interest, 3
    of Intervention, 3
Warsaw Pact, 1
Water Wars, 5
Weapons of Mass Destruction (WMD), 19
    Threats, 123, 127-29, 135, 137
Wei Qi, 142-44, 154
West, 19, 62
West Asia, 3, 58, 99, 114-15, 177, 186, 188
West Bank, 124, 127, 130
Westphalian nation-state, 3
Whole-of-government, 16

William Olsen, 15
Williamson Murray, 131
World Health Organization (WHO), 6
World Trade Center, 2, 110
World Wars, 20
*wuwei*, 145
*Wu-Zi*, 144

Xi Jinping, 155
Xiangsui, Wang, 18
*Xinhua*, 147, 160

Yaalon, Moshe, 135
Yanukovych, Viktor, 64-65, 71
Yassin, Sheik Ahmed, 126
Yemen, 99, 104, 111, 113, 185, 188-89
    Air Force, 105
    Armed Forces, 105, 107, 109, 111
    Army, 108
    Conflict, 104-10, 112
    Navy, 105
Yom Kippur War, 123-24
Yousef, Ramzi, 10
Yusuf Qaradawi, 114

Zaidi, 110
Zakat, 110
Zaldostanov, Alexander, 65
Zemin, Jiang, 146
Zuhair al-Qaissi, 126